Machine Voices

Machine Voices

A Sonic Study of Artificial Intelligence

Edward B. Kang

The MIT Press
Cambridge, Massachusetts
London, England

The MIT Press
Massachusetts Institute of Technology
77 Massachusetts Avenue
Cambridge, MA 02139
mitpress.mit.edu

The MIT Press would like to thank the anonymous peer reviewers who provided comments on drafts of this book. The generous work of academic experts is essential for establishing the authority and quality of our publications. We acknowledge with gratitude the contributions of these otherwise uncredited readers.

This book was set in Stone Serif and Stone Sans by Westchester Publishing Services. Printed and bound in the United States of America.

Library of Congress Cataloging-in-Publication Data

Names: Kang, Edward B. author
Title: Machine voices : a sonic study of artificial intelligence / Edward B. Kang.
Description: Cambridge, Massachusetts : The MIT Press, [2026] | Includes bibliographical references and index.
Identifiers: LCCN 2026001273 (print) | LCCN 2026001274 (ebook) | ISBN 9780262054720 paperback | ISBN 9780262055154 pdf | ISBN 9780262055161 epub
Subjects: LCSH: Voice computing | Intelligent personal assistants (Computer software) | Human-computer interaction
Classification: LCC QA76.9.V65 K36 2026 (print) | LCC QA76.9.V65 (ebook)
LC record available at https://lccn.loc.gov/2026001273
LC ebook record available at https://lccn.loc.gov/2026001274

EU Authorised Representative: Easy Access System Europe, Mustamäe tee 50, 10621 Tallinn, Estonia | Email: gpsr.requests@easproject.com

For Amy and Lani
For Anne and Gilbert

Contents

Acknowledgments

Machine Voices is the product of the curiosity, kindness, intelligence, and care of my family, friends, mentors, colleagues, and students—to all of whom I dedicate this work.

No one has read more drafts of this book than Larry Gross. I am deeply indebted to him for his generosity, kindness, and wisdom. Larry has inspired me in ways he would never expect. I hope one day to be for my students, colleagues, and the intellectual circles of which I am a part what he has been for me and so many others.

This project began at the Annenberg School for Communication at the University of Southern California. Here, Josh Kun inspired me to stay close to sound. Mike Ananny taught me how to study technologies and the sciences. And Kate Crawford showed me what critical scholarship of artificial intelligence could be. I am also grateful to Henry Jenkins, Tara McPherson, Jennifer Petersen, Taj Robeson, Tok Thompson, and Cristina Visperas for being incredible mentors and teachers. Finally, I am lucky to have shared time with Ally Arrieta, Herbert Chang, Sukyoung Choi, Franny Cory, Jessica Hatrick, Ana Howe-Bukowski, Simogne Hudson, David Jeong, Donna Kim, Steffie Kim, Will Orr, Chris Persaud, Tyler Quick, Cerianne Robertson, Hamsini Sridharan, and Sulafa Zidani, all of whom I am now blessed to call my friends.

The Department of Media, Culture, and Communication at New York University is where this book took its current form. Since I arrived here a little over two years ago, I've had the incredible privilege to think with remarkable colleagues and students, all of whom have pushed me in ways that I never would have imagined prior. Sue Murray helped make New York City home. Nick Mirzoeff ensured I had ample time and space to steer this book across the finish line. Alex Galloway gave me supportive feedback on an early draft of the introduction. Mara Mills showed me the ins and outs of grant writing. And Laine Nooney made key reading recommendations. My gratitude also extends to the rest of my colleagues, who have not only been so warm and welcoming but also set an inspiringly high standard in their work for me to aspire to. To the undergraduate and graduate students who took and continue to take my "AI and Society" and "Data and Society" seminars, thank you. I learn so much from you all. Finally, this project would have been impossible without the work of those who maintain and make possible the spaces we occupy. I thank all the staff of the spaces I passed through—especially those at NYU MCC and USC Annenberg—who allowed me to do my job.

Parts of this book have been adapted from previous publications. I thank Sergio Sismondo, the editor of the journal *Social Studies of Science,* as well as the anonymous reviewers who provided early feedback on "Biometric imaginaries: Formatting voice, body, identity, to data," from which I drew for chapter 1. I thank Matthew Zook, the editor of the journal *Big Data & Society,* as well as the anonymous reviewers who provided comments for "Ground truth tracings (GTT): On the epistemic limits of machine learning," from which I adapted chapter 2. Finally, I appreciate the time of the anonymous reviewers at the Association for Computing Machinery's Fairness, Accountability, and Transparency in Machine Learning conference in 2023, who provided comments for "On the praxes and politics of AI speech emotion recognition," from which I adapted chapter 3.

Mentors, colleagues, and friends unaffiliated with the communities above also graciously offered their time and wisdom. John Cheney-Lippold always provided the best support and advice when I was stuck. Juanis Becerra Sandoval has been the most incredible collaborator—key parts of this book were refined through our conversations. And Kevin Hong, my oldest friend, graciously agreed to be my "nonacademic reader" benchmark. Thank you also to the members of our "Machine Listening in the Age of AI" working group, as well as those I met through the *Knowing Machines* research collective. The conversations we shared have made their way into this book in important ways.

I'm honored to be publishing this book with the MIT Press. Thank you to Justin Kehoe and Suraiya Jetha, both of whom believed in and advocated for this project, helping me steer it to completion. I am also grateful to the four anonymous reviewers who generously took the time to read and provide comments on earlier drafts. The book benefited greatly from your suggestions. Thank you also to the copyeditors and production team who helped mold the book into its final form.

My parents' unconditional love and trust made me possible. I would not be the person I am today without their support and care. Amy Lee is my best friend, life partner, and soulmate. You make nothing else matter and everything matter more. This book would not have been possible without you. Joy and Boba, I wish you'd live forever. And Lani, I can't wait to meet you.

Introduction: Machine Voices

Voices in the Machine

This book is about voices in machines and how they get there. It parses the relations between machine, voice, and human, charting how each is invoked as a metaphor for the other, as well as the work required to build and sustain those metaphors in the context of artificial intelligence (AI).

Three familiar tropes are relevant here. The first centers the allure and fear of the machine as a substitute for the human, a story that is part and parcel of any tale of technological progress. The second invokes an age-old belief that voices are signals of humanity and the perturbation we feel when that belief is marred. The third concerns the centrality of humans and the strategic concealment of their labor in constructing spectacles of technology. This is a recurring motif throughout histories of automated machines, conjoining the automata of eighteenth-century France, cybernetics research of mid-twentieth-century Cold War America, and the veiled global network of human annotators in the development of artificial intelligence today. We will trace how these tropes clash and blend together, in recognizable and novel ways, as human voices are reimagined as mechanisms to be technologically managed by actors in the artificial intelligence industries.

To trace these themes, however, we must address a more foundational question: What does it mean to know a voice, and how is that knowledge troubled through attempts to measure, analyze, and generate it? For centuries, the human voice has been a fascination of philosophers, artists, scientists, engineers, media producers, and even politicians, all of whom have tried to tame it in their own ways using various tools and techniques at their disposal. Music historian Bonnie Gordon (2023), for instance, looks to sixteenth-century Rome to posit the castrati as "early modern cyborgs" (p. 6): surgically constructed *voice machines* to be "acquired by princes and cardinals for the perpetuation of dynastic glory" (p. 13). In the early twentieth century, English playwright George Bernard Shaw (1916/1994) wrote *Pygmalion*, a tale of social mobility in which a common flower girl, with the help of a linguist and his mechanical tools, is taken to be a duchess by training her voice to sound like one.[1] Today, artificial intelligence systems appear to have replaced such linguists and their contraptions. We now have AI voice modulation software that modifies the accents of call center workers to conform to "Standard American English" (Santos, 2023), as well as AI translation systems that enable politicians to "speak" in languages they've never previously uttered (Izaguirre, 2023).

Stories such as these have taught us to believe that the voice is a palpably knowable—and thus controllable—thing. Hypnotized by the intimacy with which it presents itself to us in our everyday lives, we have come to treat the voice as a steady site of technological intervention, especially when the right mechanical tools are available to us. But as French film theorist Michel Chion (1982/1999) reflects, the voice is anything but stable; it is a precarious and "strange object" (p. 1). Beyond the body from which it emanates, or the words, songs, and timbres through which it is heard, any attempt to define the voice is frustratingly deceptive. It's like grasping for air. We know it's there, but it averts containment. And to contain it also always means to transform it.

This book charts the contours of one such process of transformation in the increasingly important but often overlooked intersection of voice and artificial intelligence. Each chapter examines how the voice must be reconstructed in the context of AI development as a set of concrete *mechanisms*. Although such reconstructions are characteristic of any scientific and/or technological engagement with the voice, the distinct technical properties of machine learning (ML), the broader cultural imaginaries around artificial intelligence, and the financial imperatives of the technology industry fundamentally redefine the voice in ways that we have not previously contended. Ultimately, I argue that "knowing" voice through AI means something specific: It is to render the voice mechanical—manageable, measurable, analyzable, replicable, and manufacturable—to make it sellable.

Voice and Artificial Intelligence

Over the past decade, and arguably longer, there has been an explosion in what I refer to in this book as "voice AI" systems: technologies that apply ML techniques to process the human voice for various forms of identification, predictive analytics, and synthetic voice generation. My definition is distinctly broader than the conventional use of the term, which typically only denotes artificial intelligence systems that *generate* synthetic speech. Here, the term includes not only generative speech systems that output synthetic vocal sounds as audio but also *machine listening*[2] systems that receive and analyze recordings of the human voice. In the simplest terms, "voice AI," as it is used in this book, denotes artificial intelligence systems designed to both "speak" and "listen" to voices.

Unbeknownst to many, these voice AI technologies are increasingly mediating crucial decisions about our lives and the societies we inhabit. They have been and continue to be implemented in high-stakes settings, spanning health care, human resources, politics, finance, and security, but they operate in ways that are hidden from and difficult to understand for those affected. This obscurity is exacerbated by the fact that the corporations developing these technologies have every incentive to keep their operations private. As such, unlike more publicized—and thus also scrutinized—AI technologies such as facial recognition or large language models (LLMs), voice AI systems have largely evaded the scrutiny of public attention and critical scholarship.

Machine Voices addresses this gap. It investigates how voice AI technologies are designed, with a focus on the economic, social, cultural, and technical forces that mediate these design choices. Analyzing technologies along these multiple axes brings forth the broader beliefs, subjectivities, and negotiations that are otherwise hidden in the presentation and adoption of these systems but are nonetheless central to how they impact the worlds in which they are deployed. Here, a central vector of analysis is my concern with a fundamental conviction that underlies the design of *all* voice AI systems: the belief that voices are knowable and thus that "other things" can be known through them.

Consider Apple's Siri or Amazon's Alexa. These everyday features on our phones and home speakers are among the most rudimentary of voice AI technologies. But they are still illustrative of a creed that posits the human voice as a knowable object. A stranger can't declare "Hey Siri" to "wake" our iPhones. Only our own voices seem to "trigger the magic." The functioning of these basic conversational assistants, like all voice AI systems, depends on "knowing" our voices.

Across its three parts, this book will trace the myriad roles that voice is recruited to serve as sites of knowledge to be harvested and as commodities to be manufactured, so long as the right combination of mechanical tools and human expertise are available to us. I will begin with the "voiceprint," an early attempt to transcribe the voice into a readable visual diagram for the purposes of identifying criminals (part I). This will be followed by a foray into the world of "listening" machines that influence major decisions about people's lives such as hiring and compensation by analyzing the tones of people's voices (part II). Finally, I will end with an exploration of "speaking" machines that promise affordable and natural-sounding synthetic voices without the "baggage" of human bodies and minds (part III), examining how the voice is made commensurate with mechanical logics of measurement, disassembly, and reconstruction.

Here, I am especially attuned to how such a conception of voice is positioned to stand in for other slippery concepts such as identity (chapter 1), competency (chapter 2), emotion (chapter 3), embodiment (chapter 4), and progress (chapter 5). In the words of media scholar Dylan Mulvin (2021), "This is the promise of proxies and the promise of creating controllable renditions of an unpredictable and unknowable world" (p. 4). What unites these stories is thus not only the framing of voice as an object to be measured, analyzed, and replicated but also how this framing enables the transformation of these otherwise immeasurable constructs into actionable metrics. Indeed, the voice as metaphor is what makes voice AI technologies *useful* across the high-stakes settings to which they are applied. In other words, this book is about the "things" used to measure voice as well as the "other things" that voice is used to stand in for—both the proxies of voice and the phenomena that voices become proxies for. Vital to both processes, I argue, are machine metaphors.

Machine Metaphors

Philosopher Catherine Malabou (2019/2021) declares that "testable intelligence will always be *machine* intelligence" (p. 49, emphasis added). Here, she is responding to how psychologists treat human intelligence as a mechanically traceable process to be measured using tests and scales. The argument is that if intelligence is presented as data amenable to experimentation and control, it can also be considered as a programmable object to be designed, edited, and taken apart. Intelligence too is made mechanical through measurement.

Malabou's use of the term "machine" (and its derivatives) is evidently not to denote a physical system literally composed of gears and motors. Rather, it more broadly alludes to a *system of mechanisms*: a structural assemblage of distinct (or distinguishable) components that work together. The crux of this framing is its emphasis on a tangible modularity. Here, "machine" implies a modular structure of separate mechanisms, wherein each of those separate mechanisms is also separately measurable, analyzable, and rearrangeable. This mechanistic view has long been central to how the sciences have disaggregated and made sense of complex systems.

Indeed, one of the defining tenets (and critiques) of Enlightenment thinking revolved around the Cartesian metaphor of the body as machine. Progenies of this metaphor have readily proliferated since that time, of which the most relevant to this book would be in histories of information processing. There, a specific part of the body, the brain, has been described as a computational machine. In the words of science and technology scholar Paul Edwards (1996), "Cybernetics, AI, and cognitive psychology relied crucially upon computers as metaphors and models for minds . . . *Cyborg discourse*, by constructing both human minds and artificial intelligences as information machines, helped to integrate people into complex technological systems" (p. 2). Indeed, humans first had to be reimagined *as* machines for humans to also be imagined as part of them.

Such reimaginings have been a familiar trope in the history of automation, at least since the public demonstrations of automata in eighteenth-century France. There, the extravaganza was not a mechanical brain but rather a mechanical duck created by the French artist and inventor, Jacques de Vaucanson. His "Canard Digérateur" was made to appear to drink water, eat kernels of grain, and even defecate. Although the automaton was later shown to be an intricate performance masterfully executed by a hidden human operator, Vaucanson claimed that he had created a mechanical body with a gastrointestinal system in the form of a chemical laboratory, supposedly replicating the "most biological" of bodily processes: digestion.

In this way, a commitment to the body-as-machine metaphor means that every part of the biological body—brains, digestive systems, and voices—also becomes replicable *in* a machine. To attempt to trace and replicate any bodily mechanism is first to conceive the body itself as a machinic system. This brings forth the second of the two foundational conjectures that this book contends: Bodies are mechanical.

Taken separately, each of these conjectures—that voices are knowable and that bodies are mechanical—is hard to crack. As mentioned earlier, voices comprise some of the most intimate parts of our lives. We identify family and friends through the quirks in their voices. We hear emotions in voices, and we'll often trust those senses over the words spoken through them or the facial expressions behind them. Intuitively, we both *know* voices and know *through* them. To argue against this feels like an unnecessarily theoretical exercise and, frankly, almost contrarian. Similarly, treating our bodies like machines composed of mechanisms to be studied and taken apart has produced astounding results with clear practical utility. This is true not only in advanced fields such as medicine, anatomy, pharmacology, toxicology, and exercise science, among numerous others, but also increasingly in our everyday lives as we benefit from steady sleep schedules, reaching a certain number of daily steps, and managing our caloric balance. Challenging these separate beliefs is a steep hill to climb because they're reinforced both intuitively and experientially throughout our lives.

The conjectures, however, have also been the source of much harm. The belief that voices are knowable and that we can know other things through them has been used to justify wrongful convictions of accented speakers (People v. King, 1968) and deny housing to Black applicants in the US (Henderson, 2001). Accents have never only been markers of an individual's geographic and cultural origin; they have also been instruments for detecting and policing boundaries of class and race (Labov, 1973). To know a voice is presumably also to know the person who produced it, and this belief has indeed been used as an effective lever for social segregation.

Similarly, but on a much larger scale, the body-as-machine metaphor was foundational to Western eugenics. Here, the body had to first and foremost be understood as a genetic assemblage to be engineered and edited through selective breeding. This not only resulted in state-sanctioned mass sterilizations of individuals deemed "unfit" across Europe and North America during the nineteenth and early twentieth centuries, but also later served as a mobilizing ideology for Hitler's Nazi Germany and the unspeakable atrocities of the Holocaust. A mechanical body implies an editable body. And the eugenicists made sure to edit in favor of certain populations over others.

Herein lies the alluring paradox of these two conjectures. The belief that we can detect feelings in the subtleties of our loved ones' voices can also be transfigured into justifications for interrogators who claim to detect lies from speech patterns and try to use such claims as supporting evidence for convicting individuals. Similarly, our experiences improving our physical health following specific mechanistic protocols can also be doctored into a defense of biological determinism, wherein our lived experiences are reduced to mere products of evolutionary and biological processes. Indeed, like all powerful mythologies, they are dangerous precisely because they are believable and, although to a much narrower extent than their proponents may assert, frankly, somewhat true.

Machine Voices does not repeat the countless intellectual critiques of these separate conjectures. Rather, it is concerned with their combined ability to enable derivative beliefs, which then permit the development of related technologies, ultimately constructing contingent worlds configured through and around them. Specifically, it examines how the two beliefs work in tandem, focusing on their synthesis across various scientific, engineering, and business communities around voice AI. Here, this synthesis is enabled by the introduction of a third intermediary conjecture: The voice is a feature of the mechanical body. Taken together, these three conjectures produce what I refer to throughout this book as the "machine voice."

The Machine Voice

The machine voice is a voice that manifests all the characteristics of a machine, such that the voice becomes measurable, observable in parts, analyzable, able to be taken apart and put back together, reconfigurable, editable, and, of course, replicable. It is transformed from a phenomenon to a system of mechanisms. As a phenomenon, the voice is akin to how French philosopher Henri Bergson (1889/2001) describes intellectual energy: as an intensity that can only be described qualitatively. In contrast, as a system of mechanisms, the voice is transformed into a size with quantifiable features and dissectible into parts.

There is, of course, a pragmatic justification for this transformation. Like Bergson and Malabou's separate critiques of psychological intelligence, it

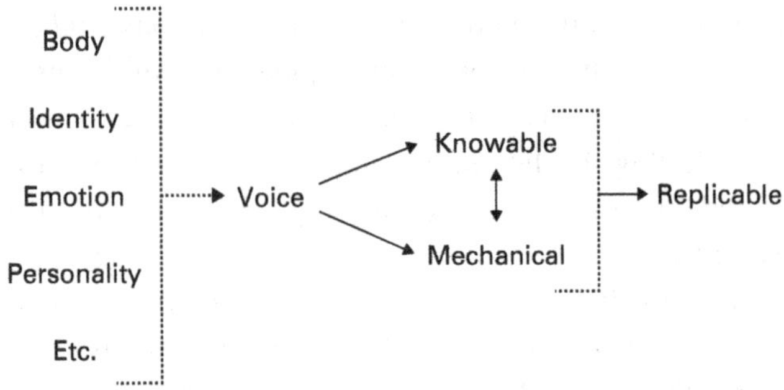

Figure 0.1
Relations between the three derivative rationales and how other constructs such as identity and emotion can be folded into this structure through the machine voice.

reconfigures the human voice into a scientifically legible and technologically engineerable object. It becomes "manageable," in every sense of the word. This manageability can be described through at least three derivative (and circular) configurations of the original conjectures produced by the machine voice (figure 0.1):

1. Voices are knowable because they are mechanical.
2. Voices are mechanical, and therefore they are knowable.
3. Voices are both knowable and mechanical, which makes them replicable.

These three derivative configurations also track the three parts of the book. Part I ("Histories and Conditions") examines the voiceprint—a spectrogram of the voice—as a manifestation of the belief that we can visualize, dissect, and measure human voices, and that such mechanistic renderings are adequate sites for ascertaining human identity. As a mechanically derived diagram, the voiceprint is presented not only as evidence for the belief but also as an actionable technology of identification. Part II ("Listening Machines") focuses on voice analytics systems that depend on treating the voice as a mechanistic object to be datafied and analyzed. Here, processes of datafication translate the voice into a measurable and actionable object used to produce knowledge about not only identity but also features related to its human speakers such as their personalities, emotions, and abilities. And part III ("Speaking Machines") examines technologies and techniques devised to engineer synthetic voices based on these same mechanistic hypotheses.

In this way, the machine voice and its derivative precepts set the stage for the technology industry, where they become especially useful for corporations selling voice AI systems. I examine, for instance, recruiting tools that claim to "solve" the "cumbersome," subjective, and expensive process of interviewing potential candidates for a job by measuring the subtle changes in their voices (chapter 2); emotion detection systems that apparently identify happier-sounding call center agents and award them with additional compensation—and vice versa for those who don't sound so happy (chapter 3); and, of course, voice generation technologies that allow media producers to do away with voice actors altogether in favor of the more flexible and financially prudent option of text-to-speech (TTS) machine learning systems (chapter 4). When phenomena are made mechanical, they become not just measurable, analyzable, and replicable but also cheaper.

Indeed, a core feature of mechanization is economization. Any story of mechanical replication also includes an economic account of competing producer–consumer dynamics, where, in this case, contestations around what voice *is* are always also in relation to how voice is *valued* and by whom. Not only that, but the mechanization of a phenomenon as multiple and economically useful as voice also renders the aforementioned "other things" that voice is recruited to stand in for as potential sites of economic activity. The machine voice thus also produces machine identities (chapter 1), machine competencies (chapter 2), machine emotions (chapter 3), machine media (chapter 4), and mechanistic understandings of progress (chapter 5), all of which can be built, sold, and purchased through more mechanistic processes. More explicitly put, they are mechanized *for the purpose* of being sold and purchased—and machine voices provide a means. This economic core means the book is as much about business as it is about science and technology. And underlying all business, science, and technology development, of course, is human labor.

Humans Making Machines

Central to this book are the humans who produce machine voices. This is meant in the broadest sense. There are humans in every part of the design, development, deployment, and distribution of voice AI systems, and they are not always (and sometimes are never) in conversation with each other. There are the usual suspects in the product developers, data scientists, computer scientists, and the machine learning engineers. But there are also those who participate little in the technical engineering but nonetheless wield incredible power in how these technologies are designed and for what purpose. These might include business strategists, senior-level administrators, C-suite executives, and venture capital (VC) investors. Moreover, the necessity of all AI systems for "quality" data at "scale" situates not only database managers and data annotators as indispensable nodes in this developmental pipeline but also the actors, singers, talent acquisition workers, and even incarcerated people enlisted to produce the data to be managed and annotated by them.

To center these humans and their day-to-day practices is to ask: Who makes the machine voice, and how do they do it? To answer that question, I conducted hundreds of hours of interviews with workers occupying these vastly different roles, as well as observational and immersive fieldwork in the physical and digital spaces that collectively comprise the voice AI enterprise. To study the machine voice is to study not only what it is or why it is useful but also how it is "made to work." As such, we will trace how humans must devise ways to mechanize certain "parts" of the voice (but not others), organized along certain taxonomies (but not others), aligned with particular formats (but not others), and to be used in some contexts (but not others). Indeed, tracking these mechanistic processes also reveals a parallel story of subjective valuations: Some things must always be made to matter more than others.

Examining these processes of subjective valuation and improvisation is also crucial to understanding why and when a voice AI system *fails*. Here, failure can be defined in several ways—technically, socially, ethically, or financially. But they all lead back to the humans designing the systems and defining what their failures (or successes) are (Passi & Sengers, 2020). This focus on human choices and practices is important because it makes the consequences of otherwise opaque machine learning operations more transparent to individuals outside of core research communities. As journalist Hilke Schellmann (2024) documents in the context of AI tools used for hiring, everyday people have limited knowledge of how these fickle technologies are designed but are nonetheless affected by them in crucial ways. Focusing on the humans who make these systems "work" (or not) can thus aid in taking concrete steps toward enforcing accountability in a space in which that is all too lacking.

Indeed, technology companies have consistently hidden behind tropes around the "black box" and algorithmic opacity (Burrell, 2016). These are themes that originally emerged as critiques against the "internal" obscurity of algorithmic processes but have since been co-opted by corporations to offload responsibility regarding the harmful consequences of their data-intensive systems (Elish, 2019). As the purview of not just voice AI but also artificial intelligence writ large expands to include applications that use machine learning for high-stakes social decision-making (e.g., border control, criminal recidivism prediction, behavior profiling, emotion analysis, loan default prediction, employee retention, and candidate success prediction, among others), it will become even more imperative to do away with this black box alibi.

Importantly, this shift toward examining human choices also allows us to parse what is and isn't an appropriate use case for machine learning. As computer scientists Arvind Narayanan and Sayash Kapoor (2024) emphasize in their exposition of "AI snake oil"—"AI that does not and cannot work as advertised" (p. 2)—a fundamental problem with many current AI technologies is not their awesome power but rather their inefficacy. When inefficacy is packaged as an affordable solution for previously expensive or nearly impossible problems and then sold to customers (including governments and other businesses) who have little knowledge of what AI can and can't do, dire consequences follow.

This is especially true in the context of voice AI technologies because, as mentioned earlier, they are overwhelmingly marketed and applied to high-stakes settings. All artificial intelligence systems (like all sciences and technologies) establish their authority through the rhetoric of neutrality, objectivity, and fairness (Porter, 1995). But numerous studies have shown that there is a constellation of social, political, cultural, and institutional forces that mediate all aspects of AI development (e.g., Crawford, 2021; Dhaliwal et al., 2024; Engdahl, 2024; Muldoon et al., 2024; Newlands, 2021; Orr & Crawford, 2024; Tubaro et al., 2020). This makes it imperative to lay bare the subjective human beliefs and choices underlying their construction.

One objective of *Machine Voices* is thus to serve as an example of critical technology scholarship that doesn't just "critique for critique's sake." Rather, it is committed to ensuring that areas in which machine learning are truly beneficial can be further explored and developed. This is only possible, however, after we've sifted through the noise. Here, we must ask what exactly these systems are designed to do and how, as well as for whom these systems are meant to work. The first step of such a process is to reject a monolithic understanding and representation of artificial intelligence. Instead, we must understand it squarely as a modifiable tool that is useful for some things, for some people, and in some situations, but not others.

"Patchwork" Methodology

In line with this commitment, a core contribution of this book is a method I introduce in chapter 2 called "ground truth tracings" (GTT). This method foregrounds the specific processes through which human developers must translate a "real-world problem" into a "machine learnable" one. Here, the aim is to make transparent the human assumptions, compromises, and ideologies that crucially mediate the development of machine leaning systems. As Florian Jaton (2021) writes, "We get the algorithms of our ground truths" (p. 81). This means to study algorithmic systems is also to study the human-designed ground truths upon which they're constructed.

I devised GTT to make sense of my own encounters with voice AI companies that presented unbelievable claims around the supposed predictions their voice analytics software could make. I found that this mixed process of deduction and empirical tracing, however, could apply beyond these contexts, and that it could be especially useful for investigating proprietary machine learning systems broadly. Indeed, a primary challenge of conducting research on machine learning technologies is not only their algorithmic opacity but also, in the case of proprietary systems, the shielded corporate contexts in which they're developed (Burrell, 2016).

In response to this secrecy, technology researchers have generally adopted what media scholar John Cheney-Lippold (2017) refers to as a "patchwork" approach (p. 32), wherein materials and empirical data from a variety of sources are patched together to better understand the stories that underlie and propel the development of these systems. GTT is one part of my broader patchwork methodology, in which I also rely on interviews, fieldwork, patent analysis, archival analysis, and critical discourse analysis to trace how voice AI systems are designed. Simply put, I study voice AI technologies by attending to what's around them.

Studying Artificial Intelligence

> Just what are we talking about when we talk about "AI"?
>
> —Lucy Suchman, "The Uncontroversial 'Thingness' of AI," 2023

I begin both my undergraduate and graduate seminars, "AI and Society," with this question. "ChatGPT," of course, is by far the most common answer. But others have also remarked, more hopefully, "the future" or, half-jokingly, "our doom." "Cheating" is another surprisingly common answer and an increasingly reliable harbinger of some of the AI-generated reading responses I read later in the semester. "A disruption in technological efficiency across multiple industry sectors" was what I believe one of my interview-ready undergraduates from the NYU Stern School of Business remarked. "Siri" and "Alexa" are also common answers, or sometimes "AlphaFold" and "DeepSeek" for those up to date on the latest AI headlines. "Consciousness" is another interesting one. Unfortunately, this feels like a symptom of the much too publicized discourses around artificial "general" intelligence (AGI), arguably one of the most contested neologisms in the field. I've also gotten "Elon Musk" more than once. Interpret that as you will.

Suchman's question—"What are we talking about when we talk about 'AI'?"—distinctly circumvents the more direct and difficult question: What is AI? Instead, it points toward an expansive network of actors that proximately congregate *around* AI. As a result, my students' responses range from well-known consumer products *branded as AI*; conditioned talking points prepared for future employers signaling that they're ready to be management consultants in industries *disrupted by AI*; public figures who like to associate themselves with the *future of AI*; existential statements on humanity and the positive or negative *impacts of AI*; and even implicit confessions to "cheating" (although I personally wouldn't call it that) on class assignments *facilitated by AI*. In this way, we talk *about* AI without necessarily defining it.

This lack of a definition is often the fate of many hackneyed buzzwords. Indeed, artificial intelligence has become increasingly impossible to bound, which means studying it has also become an exercise of defining its boundaries. Attending to what's *around* AI is thus not only a savvy way to circumvent proprietary systems or a righteous campaign to surface the humans who make them work but also a required maneuver for tracing the borders of our system of study.

For instance, AI cannot just be distilled into a set of mathematical operations. There are one too many creative human choices that distinguish an effective machine learning model from a less effective one (Domingos, 2012). AI is also not simply a human network of machine learning engineers and annotators. Their strategic invisibility in dominant tech narratives (Newlands, 2021) and frequent exploitation in the case of lower-level annotators (Williams et al., 2022) should make that abundantly clear. To describe AI as human intelligence in a machine is also wholly inaccurate. Human intelligence itself lacks a clear scientific definition, and even the ones provided by the fiercely empirical psychometrists would not conform to the properties of existing AI systems (Chollet, 2019; Gignac & Szodorai, 2024). In fact, to refer to AI as "it" also feels silly. "It" suggests a kind of unitary cohesion. But as Narayanan and Kapoor (2024) jest, this is like lumping all transportation mediums, spanning bicycles, spaceships, automobiles, and boats, under the general term "vehicles" (p. 1). Unlike the many words for different vehicles, however, "AI"—*it*—is the only term we currently have, so we must do the extra work of charting its perimeters.

AI scholar Kate Crawford (2021) tackles the question by declaring what AI isn't: "neither *artificial* nor *intelligent*" (p. 8). So, what is it? Crawford suggests that it resembles a "registry of power," pointing to the planetary flows of capital and dominant ideologies through which AI is made possible (p. 8). True. But it's not only a registry of power; it's also an echo of beliefs about what humans are and can do.

One rare consistency of all artificial intelligence systems is that they are positioned as substitutes for humans in some capacity. Why else would we be so anxious about their replacing our labor? Even the most optimistic, such as Tim O'Reilly, pronounce that "AI is a mirror, not a master" (Press, 2017). This mirror-of-humanity analogy is most often invoked to describe AI systems as products of their human-produced training data. The analogy, however, is also apt to describe how AI systems are organized along "modalities" that mirror fundamental human senses and features such as "vision" and "language."

Historically and into the present, computer vision and natural language processing (NLP) have undeniably been the two most studied and developed modalities in artificial intelligence research. Recall: ChatGPT, Dall-E, Sora, Midjourney, Stable Diffusion, Claude, Gemini, BERT, LLAMA, and Clearview AI. All of these highly publicized machine learning systems, developed by the world's most deeply funded AI labs such as Microsoft's OpenAI and Google Deepmind, fit squarely into one of those two domains. Central to these efforts, of course, are beliefs about "how humans see" and "what language is." Some critical scholars, for instance, have critiqued that computer vision is "rooted in a decontextualized and a non-situated, physicalist account of human vision" (Denton et al., 2021, p. 6). Similarly, LLMs have been famously compared with "stochastic parrots," contrasting their statistical prediction of words with the way humans construct and communicate meaning through language (Bender et al., 2021).

Indeed, we have witnessed a deluge of critical scholarship (e.g., Ananny & Crawford, 2016; Bender et al., 2021; Buolamwini & Gebru 2018; Crawford & Paglen, 2019; Denton et al., 2021; Jaton, 2021; Mackenzie, 2017; Raji et al., 2021; Raley & Rhee, 2023; Roberge & Castelle, 2021; Schulz & Dincer, 2022) that contends with the ways human vision and language have been (problematically) "mechanized"—to use the terminology of this book. As a result, we now know a lot more about both the capabilities and limitations of machine vision and machine language in artificial intelligence development; we have studied (and continue to study) what's around them. We still don't know very much, however, about machine voices.

The "things" around voice AI systems are not entirely like the things that orbit other AI technologies. Voice AI also uniquely denotes a world of sounds. As we will read, it *imagines* a world of sonic signals that reveal previously inaccessible knowledge around human identities, emotions, personalities, and bodies "trapped" in machine voices waiting to be amplified by human engineers and their mechanical tools. Unlike vision, voices engender material sites to be mined and harvested. Unlike language, voices are believed to stem directly from our bodies, not only our minds. To study a world of machine voices is thus to take a path distinct from those taken in other critical AI scholarship. It is to embark on a *sonic study*.

A Sonic Study

In his introduction to *The Sound Studies Reader*, communication scholar Jonathan Sterne (2012) writes that sound studies "redescribes what sound does in the human world, and what humans do in the sonic world" (p. 2). The key here, as Sterne himself also points out, is "redescribe." Sound studies is not about describing what we already know or believe about sound. Rather, it sets out to trouble the sonic knowledge we take for granted.

This book is indeed about testing everyday sonic categories. We will have to redescribe the act of *listening*, for instance, as we lend machines the authority to also engage in it. This is to ask not only "is machine listening listening?" (Sterne, 2022) but also how machines are made to learn to listen and why some believe this to be desirable. Answering these questions requires us to trace what we generally believe listening is and further examine which aspects of it are deemed meaningful as it is transformed into a mechanistic process. In this way, a study of machine voices is also a study of the operations through which they are made mechanically audible *as* voices.

As will be evident throughout the book, however, machine listening is anything but a unitary practice. It can refer to the process of transcribing the sounds of voices into visual diagrams, in which listening is translated into an analysis of charts and a visual tracing of fluctuating frequencies. It can also refer to a specific type of listening that humans engage in, where they must follow rigid mechanistic protocols for how (and how not) to listen and what (and what not) to listen for. Anthropologist Beth Semel (2022) describes this as "listening like a computer" (p. 267). And listen like a computer we must if we are to teach computers how to listen like us—or at least that is the belief held in some machine learning communities and technology companies. Machine listening also orients us toward a reevaluation of our own listening practices. As Sterne (2022) writes, "Human listening is already technological" (p. 3). If so, it appears necessary to parse how machine listening is *differently* technological.

By testing the boundaries and mechanical forms of listening as such, this book also divulges the mechanical "parts" of voice that are produced through them. Indeed, the machine voice does not have *a sound*; rather, it depends on *a collection of many sounds*. Synthetic structures are key features of mechanization. So, when the voice is turned mechanical, it is also retroactively imagined as a synthesis of various mechanical components. In turn, this renders it reconstructible through a combination of those various mechanisms. As we will read in parts II and III, such synthetic logics are central to how both AI machine listening and speech generation operate.

Specifically in chapters 3–5, we will read about massive datasets containing millions of audio files that underlie the construction of voice AI systems. These audio files contain recordings of vocal utterances by intentionally varied speakers evaluated by similarly diverse listeners, uttering and listening based on predefined mechanistic guidelines. Here, both machine listening and machine voices emerge as synthetic aggregates of these mechanically mediated sonic practices. Indeed, unlike the intimacy and situatedness with which we associate human voices, machine voices are, by design, distributed and multiple. They are not concerned with the oddities of a particular voice.[3] Rather, massive datasets are compiled precisely to do away with those particularities. The goal is instead to identify general patterns *across* speakers to derive vocal "types" or "norms." This means that to study these statistically derived machine voices is also to redescribe them in mechanistic terms—as multiple, synthetic, and modular.

To be sure, sound studies as a field already treats the human voice as a manifold concept. Voice scholars Nina Sun Eidsheim and Katherine Meizel (2019) write, "Voice and its surrounding values represent no stable category, rather one that is negotiated with each utterance and each listening" (p. xiv). Sonic studies of voice have thus inevitably invoked its multiple *ontologies* and *epistemologies*. In simpler terms, these refer to the different forms that voice appears as and the different methods through which it is made knowable. For instance, the voice in opera is literally a different "thing" from the voice in speech signal processing. Each has its own objectives, practices, beliefs, and tools about what to do with voice and how.

The voice AI enterprise and the machine voices at its core come with their own such conventions. To frame this book as a "sonic study" is thus to examine how voice appears in action in the context of voice AI technologies. It is to ask "What does the voice need to be?" and "How does it become those things?" Inevitably, these questions also point to the various forces at play—social, cultural, political, economic, and technological. These forces are not always equally assertive. They are in flux, but they are always present nonetheless. And they collectively mediate how the voice AI enterprise manages what voice is and which parts of it are useful.

As "a sonic study of artificial intelligence," then, *Machine Voices* troubles and redescribes the sonic beliefs and practices that orbit voice AI technologies. It resists the paralysis that stems from the ambiguity of the term "artificial intelligence" by positioning it as a testable hypothesis of human capability. Similar to how computer vision models inherit a belief of "how humans see," or how LLMs employ a proposition about "what language is," voice AI systems invoke a thesis of how humans listen and speak. Throughout the book, I show how this thesis of mechanization is constructed by tracing the concrete ways in which players in the voice AI industry massage and transform voices to fit the mechanical formats required for artificial intelligence. Here, my position is deceptively simple: Machine voices are not human voices. This is the point. And throughout the five chapters of this book, we will learn that there are crucial trade-offs, should we forget.

Chapter Outline

Chapter 1 starts before the time of voice AI, with an audit commissioned by the United States Federal Bureau of Investigation (FBI). The technology in question was the use of speech spectrographs for identifying speakers. As a wartime method that was experimented with in the 1940s, so-called voiceprinting was facing a crisis by the 1960s, as scientists repeatedly challenged its use in court. The premise that a visual transcription of an audio recording of voice can be reliably presented as evidence simply could not be backed by science. Despite this, voiceprinting technologies reemerged in the 2000s under the name "automatic speaker recognition" (ASR). Touting advances in machine learning as ways to circumvent the previous shortcomings, technology companies reframed the problems of voiceprinting as ones of engineering, statistics, and data—which machine learning is well positioned to address—and not one of science and theory. I show how such a framing sets the stage for the ensuing beliefs that propel the voice AI industry.

In chapter 2, I turn to the world of "listening" machines that promise to tell us about who we are and what we're capable of by analyzing the sounds of our voices. Here, I test the claims of a voice AI startup that declares its AI technology can determine how well a potential job candidate will fit into a team. But to evaluate this bold claim, we must ask: How is a machine made to learn to listen? Machine listening is different from human listening, so we must devise new methods to trace it. In response, I introduce a framework called "ground truth tracings," an approach that foregrounds the human decisions required to translate a qualitative problem into a quantitative one. It turns out that machines are made to listen through a sequence of human decisions that privilege some scientific theories over others, some ears over others, and some methods of engineering over others.

Chapter 3 delves deeper and examines this process in the context of emotion. Specifically, I examine AI speech emotion recognition (SER) technologies, which claim to detect emotional signals in our speech. This claim, however, assumes that we know what emotion is. The simple answer is that we don't, at least not scientifically. There is no scientific consensus on how to define emotion, let alone how to measure it. I review different theories of emotion and show how some are more compatible than others with the practical constraints of machine learning, ultimately tracing how these theories get entrenched in the datasets that underlie the development of SER systems. I argue that this unwittingly results in a concentration of power among the creators of those datasets, effectively rendering them *emotional designers*. When SER technologies are deployed as a means of evaluating and compensating workers based on how "happy" they sound, such design choices matter.

In chapter 4, I shift my focus to "speaking" machines. Like most technologies, these systems are packaged in rhetorics of democracy, promising the dismantling of language barriers and the wide availability of realistic synthetic voice libraries for content creators. We must remember, however, that the original masters of vocal manipulation are voice actors. Here, I give some space back to them. As we will read, voice actors are recruited to produce training data for the development of synthetic voice generation systems. Indeed, synthetic voices depend on the bodies and expertise of voice actors but ultimately serve to replace them. By recounting my conversations with both voice actors and tech workers, I contrast their divergent approaches to voice generation: artistic and bodily versus statistical and machinic. In so doing, I reflect on how these divergent approaches mediate our understanding of what voice can be.

In chapter 5, I confront the business of AI voice generation. Here, discussions of scientific progress, financial profitability, and how voice can be made useful converge in the context of machine learning benchmarks. Benchmark datasets are the central artifacts through which technical performance and progress are communicated in machine learning. But gold-standard benchmark datasets don't exist in the voice AI community to the same extent that they do in computer vision or NLP. Not only that, but benchmarking as a practice has been widely critiqued as an inadequate method for understanding the utility and performance of machine learning systems in real-world applications (e.g., Engdahl, 2024; Orr & Kang, 2024; Raji et al., 2021). In turn, I examine how economic benchmarks replace technical ones in the world of machine voices by following the decision-making processes through which venture capitalists justify hundreds of millions of dollars in investments. Here, scientific progress is reframed as economic viability—a shift that I argue introduces new concerns beyond the narrow technicalities of traditional ML benchmarking.

Finally, I move beyond the context of voice AI in the conclusion and reflect on how the stories in this book can together inform our thinking around the ultimate metaphor underlying all artificial intelligence development: the human.

* * *

Machine Voices seeks to expand our collective knowledge of both sound and artificial intelligence. Sonically, it pushes the boundaries of voice and listening as both become targets of machine learning and statistical knowing in ways that previous analog and digital sound technologies have not contended. From the purview of artificial intelligence, it structures new grammars for understanding sonic modalities amid a body of critical literature overwhelmingly occupied with vision and language.

These interventions are especially urgent in the current moment. We are at the cusp of a cascade of changes that machine listening systems and synthetic voice AI technologies will bring forth as they become central parts of our media, economic, and government infrastructures. But on an even more foundational level, fundamental understandings of humanity, intelligence, perception, and truth are being renegotiated as we speak. This is not meant in an abstract sense. Rather, the group of technologies, technologists, and owners of capital who reign bearing the flag of "artificial intelligence" are redefining vital categories through which we have come to understand ourselves and the worlds we live in. Dedicated efforts will thus become increasingly important for not only studying the contours of these emergent categories but also resisting them by pointing out their contradictions and failures. In the words of science and technology scholar Donna Haraway (2016), we must "stay with the trouble." And stay with the trouble we will.

I Histories and Conditions

1 Frames of Voice

An Empirical Art

In March 1976, the FBI commissioned a report from the National Academy of Sciences to evaluate whether speech spectrograms could be used to reliably identify speakers from the sound of their voices. The request was a response to more than a decade of courtroom controversies in which so-called voiceprints—spectrograms of vocal sounds—were presented as evidence for identifying and convicting individuals. In response to the FBI's request, the United States National Research Council appointed the Committee on Evaluation of Sound Spectrograms, a group of nine professors and researchers from fields spanning physics, psychology, linguistics, audio engineering, audiology, speech pathology, and law, to lead the investigation.

Three years later in 1979, the committee published a 161-page report titled "On the Theory and Practice of Voice Identification." It stated the following:

> The Committee concludes that the technical uncertainties concerning the present practice of voice identification are so great as to require that forensic applications be approached with great caution. The Committee takes no position for or against the forensic use of the aural-visual method of voice identification, but recommends that if it is used in testimony, then the limitations of the method should be clearly and thoroughly explained to the fact finder, whether judge or jury. (National Research Council, 1979, p. 2)

The verdict was based on several factors. To begin, the committee suggested the use of the term "voicegram" over "voiceprint," asserting that the latter misleadingly equates speech spectrograms to fingerprints. Unlike a fingerprint, "a direct graphical representation of anatomical features, the ridges in skin," a speech spectrogram is an "indirect representation of a complex process that produces vocal sounds" (p. 6). This complex process, according to the committee, includes not only anatomical relationships but also environmental factors related to physiology, psychology, and age that can drastically affect phrases uttered by the same speaker differently. Not only that, but the mechanical system that "detects, transmits, records, and prints the voice information can affect the final graphical picture in major ways" (p. 6).

Indeed, the voicegram could only amount to an aggregate representation of heavily mediated and indistinct factors. For the committee, this rendered it significantly less reliable than the explicit inscriptions produced through more direct biometric techniques such as fingerprinting. Furthermore, the lack of scientific research that could demonstrate how and to what extent each mediating factor (technology/equipment, physical environment, psychological state, context, etc.) would affect the resulting graphical representation crucially undermined the voicegram as a reliable tool for identification.

In addition to the variable conditions mediating how voicegrams were produced, the committee also recognized that the successful *use* of voicegrams for identifying speakers was inextricably tied to "the skill of the examiner making the judgments" (National Research Council, 1979, p. 60). The practice was closer to a craft to be honed through training and experience than a systematic technique to be replicated following clear mechanistic protocols. The "skill and judgment of the voicegram examiner . . . developed mainly through extensive experience in comparing voicegrams" (p. 14) were inevitable elements of the operation. But as the committee argued, "So much experience would not be needed if explicit and objective criteria were available" (p. 14).

Based on these limitations, the committee made four recommendations:

1. Extensive research to establish a basic scientific understanding of voice identification.
2. Development of objective standards and methods for testing the performance of voice identification examiners and a system for certifying their competence.
3. Improvement in the methods employed for voice identification practices and the training of voice examiners.

4. Transparent communication of inherent limitations of voicegrams if they are used as evidence in court to protect against the overvaluation of such evidence.

The committee was explicit in singling out the first recommendation of establishing a "basic and comprehensive understanding of the scientific aspects of voice identification" (National Research Council, 1979, p. 63) as the most significant long-term goal. It grouped the remaining three as a more pragmatic and short-term "call for actions that could lead promptly to improvements in the present practice of voice identification and its use in forensic applications" (p. 63). The report thus acknowledged the underlying *potential* of using voicegrams for the mechanical identification of voices, given certain tightly scoped conditions. But it ultimately concluded that considerable research in both science and engineering was still required to reliably use it beyond controlled laboratory experiments. In the committee's words, it was closer to "an empirical art" (p. 1) than a "fully developed technology based on science" (p. 11).

Managing Voice

The voice in the voicegram was thus still unruly in 1979. As the committee's report details, the unruliness was primarily the product of a dearth of scientific knowledge: We simply did not know "what aspects of voices point most directly and accurately to the identity of the speaker" (National Research Council, 1979, p. 15). Histories of science and engineering, however, have taught us that "complete" scientific knowledge is neither always possible nor necessary to make technologies "work." The committee itself acknowledged that although the problem could not be adequately addressed without further scientific research, it could still probably "be dealt with in a purely statistical manner, without consideration for the physical sources of speech sounds and their variability" (p. 10). For the committee, statistics could potentially serve as a temporary fix in the absence of robust scientific theory.

Extensive scholarly efforts have examined the role of statistics in managing societies' uncertainties (Desrosières, 1993/1998; Foucault, 1966/1994; Hacking, 1981; Porter, 1995). Often grouped under the banner of "the sociology of quantification," this body of scholarship has examined topics ranging standardization protocols, bureaucratic governance, sociopolitical decision-making, and the quantified self as practices that emerge out of a broader social epistemology that privileges numerical thinking. A central tenet of this epistemic frame is the assumption that social phenomena follow mechanistic principles, wherein individual, social, political, and even cultural behaviors can be reimagined as patterns to be represented through numbers and parsed using statistical means. As many scholars have observed, however, the belief in such underlying mechanistic patterns has not necessarily manifested as a formal hypothesis to be rigorously tested through the scientific method. Rather, it has been more effective as a galvanizing force for governing bodies to act strategically based on the assumption that such patterns exist (Porter, 1995). The objective is less to understand why phenomena occur (science) and more to decide how to move forward, given past and current conditions (*data* science). Simply put, the currency shifts from evidence to action, wherein the ability to make use of and act on data is privileged over the identification of causal relations.

From Body to Image to Number

This chapter uses the voicegram as an entry point to examine how voices were similarly manipulated into actionable and ready sites for extracting knowledge about their speakers through mechanical means. Specifically, it traces how understanding the voice was reframed as a project of *managing its unruliness*—that is, to rearrange the voice into stable structures compatible with mechanical frames and tools of analysis. To examine how the voice was made manageable in this way, I chart three relevant paths.

The first concerns histories of how the voice came to be seen as a mechanism of the biological body long before the advent of spectrographic technologies. Across diverse fields such as medicine, psychology, and singing, the human voice was positioned as a sonic signal / product of inner bodily workings. It became both an entry point for *diagnosing* the body, in the way a mechanic attends to creaks and squeaks in a mechanical system, as well as a product refined through *training* the body, akin to how a musician polishes and tunes a musical instrument. Seen as bodily signals, voices could be rendered underlying mechanisms of the body to be accessed through any available (i.e., non-sonic) medium. This was a key frame of voice that rendered it accessible, flexible, and manageable in the context of speaker recognition technologies.

The second path charts how human "experts" emerged as central actors in materializing such frames of voice in the development and use of speaker recognition systems. Examining the contested figure of Lawrence Kersta, a core actor in the popularization of voiceprinting in the mid-twentieth century, as well as the emergence of BATVOX, a widely used ASR technology that emerged in the early 2000s, I parse the relations between expertise and objectivity as speaker recognition technologies transitioned from spectrographic (visual) to statistical (numerical) frames.

The final trajectory traces how the mechanization of voice and its applications gave way to mechanical solutions. When voice is framed as a stable mechanistic process to be traced, the task of speaker recognition is also transformed from a theoretical investigation of *if* voice can serve as a reliable marker of identity to an engineering challenge of *how* voice can serve as a reliable marker of identity. Tracing developments in speech signal processing and deep learning, I examine how statistics and machine learning ultimately replace the scientific method as the primary instruments of epistemic authority in speaker recognition technologies.

The convergence of these three paths serves as a necessary backdrop against which the rest of the stories of this book will unfold. For voice to be rendered a viable domain of artificial intelligence, frames of voice must be established in which the voice can be treated as a manageable mechanism to be traced, analyzed, edited, and replicated. Here, we will chart one such history of how the voice was made commensurate with these mechanical frames.

Sound Writers

Communication scholar Xiaochang Li and historian of science Mara Mills (2019) write that "engineers and anthropologists began collecting images of human voices almost as soon as graphic inscription tools such as the phonautograph and logograph became available" (p. S132). Originally, these tools were used for the visualization of *speech*: the parts of vocalization that pertain to linguistic communication. Proposed for uses in phonetics and deaf education for "teaching the deaf and mute to speak as if they could hear" (Sterne, 2003, p. 36), these early "sound writers" were seen as a type of prosthesis that could assimilate the communicatively marginalized into more "standard" arrangements of spoken communication: "The device would allow deaf people to *see* the sounds that they were making with their voices, thereby allowing them to modulate the sounds they made until they matched the tracings of vowels or consonants spoken by a hearing person" (p. 38). The objective was not to identify the particularities of an individual speaker but rather to uncover the commonalities of a "vowel" or "consonant" across multiple speakers. The visual recognition and matching of graphical patterns supplanted the need to "hear" and auditorily process speech signals. By thus automatically transcribing the sound of voice onto a piece of paper as lines to be traced and compared, sound spectrographs reclaimed the voice, and everything it "carried," as not only effects to be replicated by deaf individuals but also suitable scientific "object[s] to be studied" (Sterne, 2003, p. 44).

As philosopher Sybille Krämer (2009) writes, "The sciences have always been dependent on the 'eye of the mind', which gives the register of the senses access to the invisible, the abstract, the hypothetical and the imaginary" (p. 1). To transpose the sound of voice onto a diagrammatic plane thus "opens up a visual as well as tactile space to operate and manipulate abstract, non-spatial entities . . . By virtue of this twofold nature, non-visual objects and theoretical entities can be spatialized, made intuitable and manipulable" (p. 9). If signals of voice could be transposed visually and stably onto a flat plane to be scrutinized, it logically followed that *anything* the voice is believed to "carry" should also be contained in that transcription.

As such, although the earliest "visible speech" machines were used for the transcription of speech signals to identify linguistic patterns *regardless of who was speaking*, a parallel development emerged shortly after to experiment with these sound writers to identify *exactly who was speaking* (Li & Mills, 2019). The possibility of visually representing "speech" also implied the possibility of visually identifying "speakers" through their voices: Both speech and speaker only needed to be translated into visual patterns to be analyzed and compared. British statistician and eugenicist Francis Galton, for instance, experimented with the logograph as early as 1877 to examine visual samples of his own voice as he imitated different accents. Seen as objects akin to "vocal fingerprints" (Gilbert, 1911), Galton was driven by a belief that anthropologists and criminologists had at the time, which viewed "graphic inscription [of voice] as evidence that humans could *not* disguise their unique voices and ethnic origins" (Li & Mills, 2019, p. S133). In this way, graphical inscriptions of the human voice were used not only to identify but also to *typify* human speakers. Depending on how one chose to look, the universality of communication, the particularity of speaker identity, and even the presumed typologies of speaker groups could all be extracted from these visualizations of the voice. Or at least that was the belief at the time.

The core question, of course, was exactly how to make those distinctions. Despite the alluring possibilities they promised, sound spectrographs (automatic sound writers) simply produced patterns of lines. These spectrographs made voices (and everything they "contained") visible, but its practitioners still needed a framework—a theory—through which to make sense of and analyze the spectrograms (visual graphs) they produced. It was a matter not just of writing voice but also of *how to read* it.

As Bernard Kamine (1969), the Deputy Attorney General of California between 1969 and 1972, wrote for the *San Diego Law Review*, the use of sound spectrographs for the purposes of speaker identification relied on two assumptions: "(1) every individual speaks so uniquely that the differences between the same utterance by two people are greater than the differences between the same utterance by one person in varied contexts; (2) the spectrogram adequately portrays that uniqueness" (p. 224). The second assumption described the technological means, but the first assumption was the scientific hypothesis through which the so-called voiceprint technique was promoted as a valid method of speaker identification. This threshold of speaker variance—believed to be relatively narrow for a single speaker and unambiguously wide across speakers—was hypothesized as stemming from the unique vocal *anatomy* of each speaking individual.

Unlike vocal sounds pertaining to speech, which were seen as products of learned behaviors meant to be universalized across speakers, the vocal sounds stemming from the body of the speaker were seen as unique and unmanipulable. As Aristotle famously asserted, language (*logos*) represented logic and reason, a projection of the rational mind, whereas the vocal sounds "around it" (*phone*) were mere vestiges of the individual body (Rancière, 2000/2004). Spectrograms were perceived along similar lines. The sounds of "vowels and consonants" versus "everything else" could be interpreted as "language" versus "body" and then further abstracted into "speech" versus "identity." As such, sound spectrographs built on this long-standing belief that voices carry both mind and body,[1] thus not only rendering speech visible but also reclaiming the body as sonically traceable.

Sounding Bodies

This idea that bodies, and especially their less accessible interior spaces, can be examined through the sounds they make is part of a much longer history of *techniques of listening* (Sterne, 2003, p. 90). Indeed, listening itself first had to be understood as a mechanistic process amenable to scientific and technical protocols.[2] Building on what sociologist Marcel Mauss (1936/1973) calls "techniques of the body" in which he describes the body as our "first and most natural technical object" (p. 75), Sterne writes that listening, viewed in this way, also "gets articulated to notions of science, reason, and rationality. Listening becomes a technical skill, a skill that can be developed and used toward instrumental ends" (p. 93).

I build on this notion to argue that the body served two core functions in the emergence of voiceprinting as a supposedly viable method of identification. On the one hand, it allowed its practitioners to lay claim to the anatomical singularity of human bodies in the same way that Alphonse Bertillon promoted anthropometry (i.e., body measurement) as a method for identifying criminals in nineteenth-century France (Kaluszynski, 2001). The singularity of the body presented the antithesis to the universality of language, which meant the "aggregate voice" represented by the spectrogram could be divided, and thus made legible and useful, along this body–language divide. But equally as important as this claim to singularity was the body's utility in making listening and its targets (i.e., voices) technically legible as mechanisms to be replicated and traced. Indeed, even before voices were made visual through various renditions of automatic sound writers in the late nineteenth and early twentieth centuries, voice itself had to be seen as *mechanistic*. The "content" of spectrograms could be analyzed visually only if one assumed that there were mechanistic patterns underlying their inscriptions *to be analyzed*.

Voiceprinting thus relied on fastening the voice to the body—or, more specifically, reclaiming the voice as a measurable signal and function of the (1) unique and (2) mechanistic (information-laden) body. In this way, the voiceprint is also part of a longer lineage of forensic-diagnostic technologies that sought to decipher the body through its sounds. For instance, Sterne (2003) elaborates that mediate auscultation—"the practice of listening to movements inside the body with the aid of an instrument" (p. 99)—brought with it a new technique of listening that reconstructed "the living body as an object of knowledge" (p. 99). Its sounds, of which voice was no exception, were thought to provide ways to access this otherwise unobservable interiority. Technologies such as the stethoscope, which provided a means of listening to those bodily sounds, helped to "exteriorize" that knowledge: "Every movement of the organs in the human thorax could be tracked by *listening* to the body with the aid of an instrument, and those movements could be rendered meaningful" (p. 102). Within this recalibrated framework of technically listening to the body, all bodily sounds, including "vocal sounds became akin to a thermometer: an indicator of the patient's inner state" (Eidsheim, 2015, p. 358). So long as a trained practitioner had access to the right mechanical tools, listening was believed to be a viable technique for extracting knowledge from the body.[3]

The claims of psychoanalysts were bolder and more explicitly focused on the voice. Framing their interpretive diagnoses of "Freudian slips" and vocal mannerisms as "merely [uncovering in voice] what was invariably there to find" (Eidsheim, 2015, p. 358), psychoanalysts situated their practice of diagnosing patients' mental states through their voices within the frame of scientific empiricism and pure observation. Their trained ears were allegedly the primary technologies required to extract information from their patients' unconscious minds. Indeed, as psychoanalyst Salman Akhtar (2012) delineates in *Psychoanalytic Listening*, psychoanalysis reconstructs both the practice of listening and its site of audition (speech) as mechanistic processes to be replicated, traced, and measured. "Listening" is divided into discrete "models," each with their own "theoretical anchors" and "technical yields." "Speech" can be codified into spoken words, silences, and postural changes, with each category further disaggregated into more granular subcategories ("Introduction").

As "techniques of listening," then, mediate auscultation and psychoanalysis also refashioned their sonic *targets*—bodily sounds and organs, voices and unconscious minds—as "technically audible mechanisms." Not only did they not reclaim listening as a mechanistically replicable process (a technique to be learned), but they also reclaimed their targets as mechanistically traceable objects (pieces to be analyzed). Seen this way, the utility of bodily sounds became bidirectional. They comprised both *signals* to be detected and *products* to be refined. The voice could be a sonic channel for accessing knowledge about the body, but the body could also be a site for refining the sound of voice.

The castrati of sixteenth-century Rome were perhaps the most explicit manifestation of this belief. Castration is widely considered one of history's most ancient and violent forms of corporal punishment, but it was not always used as an explicit technique to produce singers. As Bonnie Gordon (2023) writes, "What made the Italian castrati different was that they were *made for their voices*" (p. 12, emphasis added). Cautiously describing the castrati as "early modern cyborgs" (p. 6) or, more specifically, "human beings turned into voice machines whose sounds were consumed" (p. 7), Gordon positions them as "pragmatic solutions to the problem of high voices . . . [at a time when] [b]oy sopranos had a limited shelf life and women were considered dangerous and excessive at best" (p. 18).

The body of the castrato thus represented a site of mechanical (surgical) intervention, wherein the objective was not to "correct" a physical attribute deemed "defective" but rather to reconstruct the body as a sonorous, strategically engineered, and culturally appropriate machine to be sold. In this way, viewing the body through a mechanistic frame not only allowed it to be explained and analyzed through its supposed mechanisms, but also enabled its reconstruction as a useful (and literal) instrument of power through the explicit manipulation of those same mechanisms.

The production of Italian castrati declined by the late eighteenth century, but the practice of training (and differentiating) singers through their anatomies endured. Manuel García II, dubbed the "Christopher Columbus of the larynx" and credited with inventing the laryngoscope (Anderson, 1905), was one of the leading figures in this regard. The 1830 colonial invasion of Algeria by the French army presented one of the earliest opportunities for García to study human anatomy as he gained access to research subjects via the Algerian military hospitals that he worked in. As a historian and theorist of music invested in vocal pedagogy, García developed an anatomical frame of vocal production, conceptualizing the trachea and "singers' differing skull shapes and sizes as the basis for their timbral differences" (Eidsheim, 2015, p. 349).

This was most likely a development from García's time studying with "French musicologist François-Joseph Fétis, [who was] known for his recourse to racial classification schemes in explaining human musical difference" (Eidsheim, 2015, p. 348). Fétis, drawing from his commitment to craniometry and racial eugenics, theorized that "voice 'sounded' the cranial dimensions" (Eidsheim, 2018, p. 40), and thus asserted that musical capacity, of which vocal quality was an indispensable component, was "the product of human faculties which are distributed unequally among peoples as well as individuals" (Fétis, 1868, quoted in Pasler, 2004, p. 26). Upholding the "Aryan" race as the most advanced—a principal tenet of any craniometric venture (Gould, 1980)—Fétis would later assert that "interbreeding with less privileged races" would have to be checked in order to prime the necessary conditions for music to progress (Fétis, 1868, quoted in Pasler, 2004, p. 26). As Eidsheim (2015) writes, "From the moment human anatomy became part of the vocal timbral equation, vocal timbre was conceptualized to reveal racial bodily essence" (p. 349).

The recruitment of voice into the "rational" frame of the mechanistic body thus also rendered it an instrument for rational*izing* similar technologies of bodily distinction such as race (Eidsheim, 2015), gender (e.g., Henrique da Cruz Martinho et al., 2024), sexuality (Sulpizio et al., 2015), and class (Labov, 1973). Indeed, this is one of the "draws" of mechanization: It makes previously uncorrelatable features compatible (and comparable) through the common frame of the mechanistic. If all such categories are rendered "features" of the mechanistic body, then our sensory faculties simply become different channels for detecting their content. "Seeing" a sound spectrogram can be deemed commensurate with "listening" to a voice if both are construed as identifying the same underlying mechanistic features that comprise it (one through lines on a piece of paper, the other through sounds). The same can be said for all that voice is believed to "carry"—speech, speaker, identity, race, gender, class, and so forth.

The readable voice thus first depended on the belief in and construction of a mechanistic and sonorous body. Indeed, if, as Sterne (2003) suggests, techniques of listening reconstructed sound "as an object in physics, acoustics, and physiology" wherein "voices or music had [previously] been privileged instances of sound, [but] now they were merely instances of a more *general category* of sound" (p. 93, emphasis added), I argue that the visual representation of voices further abstracted them into the common frame of the mechanistic body as general *signs* (figure 1.1).

Voices were no longer even just an instance of sound; they became mechanisms of the body that could be expressed (and therefore also traced and manipulated) through *any* available medium (sound or not). "Techniques of listening" could thus be made commensurate with "techniques of reading," so long as *how* one listened (or read) conformed to the underlying

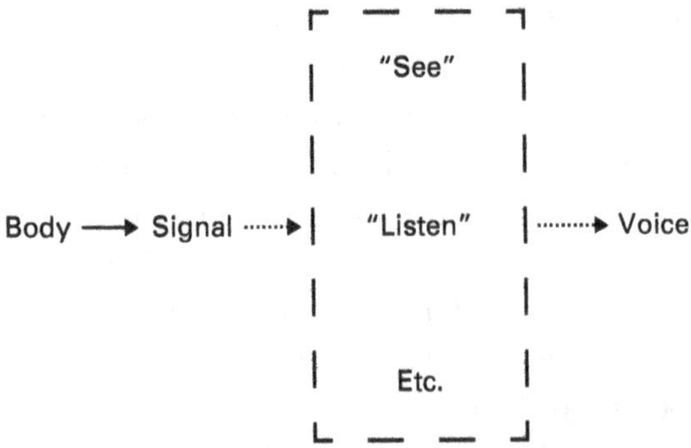

Figure 1.1
The voice as a signal of the body and how such a conception enables other techniques for attending to it.

assumption that what was being attended to were mere signs of the mechanistic body. As such, *technicians* who upheld and adhered to those beliefs became central for attending to voice in this way.

The Trap of Expertise

Voiceprinting officially became a widespread forensic method of speaker identification in the 1960s.[4] Although the practice of speaker identification using spectrographs and other automatic sound writers had been experimented with since the nineteenth century based on the idea that voices transmit bodily signals, it did not catch on as an established technique until Lawrence Kersta, a self-proclaimed voiceprint expert, entered the scene.[5]

Asserting that the shape and movements of all individual vocal tracts are unique, Kersta (1962) promoted voiceprinting via spectrographs as "analogous to fingerprint identification" and "an important aid in helping law enforcement authorities identify criminals" through their voices (p. 1253). For Kersta, visually analyzing speech spectrograms was not only *comparable* to listening to voices but also *more objective and accurate*. As Kamine (1969) describes of the underlying premise of Kersta's voiceprint technique, "The ear distorts certain sounds; the spectrograph does not. Therefore, visual spectrogram comparisons should be more accurate than the ear for identification" (p. 224). Indeed, if the voice was believed to be a signal of the mechanistic body that could just as effectively be traced through graphic inscriptions as sound, comparing stable spectrograms would surely comprise a more objective method of analysis than simply listening with one's ears.

In addition to promoting this belief, however, a significant part of Kersta's campaign also depended on treading the line between the necessity of his "expertise" in the voiceprint technique and the "mechanical objectivity" (Daston & Galison, 1992)—that is, replicability—of the method. As historian of science Theodore Porter (1995) argues, this is a common tension in the pursuit (and construction) of objectivity. "Expert knowledge is almost by definition possessed by only a few, and no such art is ever reduced to a handful of rules that can be looked up and mastered by anyone with a textbook" (p. 7). Megill (1994) describes this type of objectivity—that which is tied to expert opinion—as "disciplinary objectivity," elaborating that it is often only accepted in certain specialized contexts such as hospitals or courtrooms. Even in those settings, however, it is contingent on "an ability to reach consensus . . . within a specialist disciplinary community" (Porter, 1995, p. 3), wherein that community must also be able to establish trust among the wider public. In the absence of such consensus and trust, disciplinary objectivity crumbles. Instead, it is replaced with the idea that "expertise should be mechanized and objectified" (p. 7) and thus distanced from any specific individual or community. Here, human experts are supplanted with a formalized set of replicable protocols that ascribes epistemic authority to standardized rules and methods. Indeed, mechanical objectivity "implies personal restraint. It means following the rules . . . [because] [r]ules are a check on subjectivity" (p. 6).[6]

This was the challenge facing Kersta. To build credibility for his supposedly objective and accurate voiceprint technique, Kersta relied on emphasizing both his own expertise and the alleged robustness of his methods through spectrographic voice identification experiments he designed and conducted. Across these experiments with "high school girls who had received a week's training in the features of the spectrogram which he thought were relevant for speaker identification" (Kamine, 1969, p. 227), Kersta (1962) reported an average accuracy rate of 99 percent across the trained subjects. He also promoted a "two week, in residence training program" offered by his company Voiceprint Laboratories, which would be followed by an additional two years of mandatory fieldwork (p. 228). At the end of this period, "when a level of proficiency [was] reached which Kersta consider[ed] adequate, the student [was] deemed an 'expert' in the technique" (p. 228, footnote 73). Under the authority and guidance of Kersta, voiceprinting appeared—at least partially—to be both a mechanically (replicable among his students) and disciplinarily (tied to his own expert opinion) "objective" method of forensic identification.

The high rates of accuracy Kersta reported in his experiments and training programs, however, could not be replicated by others. This justifiably raised serious concerns from across relevant scientific communities. Outside of one study by researchers from Michigan State University (MSU) that "confirmed the figures reported by Kersta in 1962" (Tosi et al., 1972, p. 2041), the majority of subsequent studies examining the replicability and scientific basis of voiceprinting challenged the claims and design choices of both Kersta and the MSU study (e.g., Black et al., 1973; Bolt et al., 1973; Hazen, 1973; Hollien, 1974; National Research Council, 1979; Reich et al., 1976).[7] Reported accuracy rates varied widely, and the relations between voice, body, and identity were still unclear. There was neither a set of formalized rules and procedures through which one could reliably replicate the purported accuracy rates of voiceprinting claimed by Kersta nor a scientific consensus on its validity. As a result, the admissibility of voiceprinting as a form of forensic evidence in court steadily declined, starting in 1974, and it was effectively disavowed after 1979.

From the perspective of establishing voiceprinting as an acceptably objective method, its eventual demise was inevitable. Its "legitimacy" could be sustained only through a commitment to Kersta, the expert *individual*, and not on either formalizing voiceprinting as a mechanistically replicable *protocol* (mechanical objectivity) or establishing it as a scientifically backed approach supported by an expert *community* (disciplinary objectivity). To repeat the words in the 1979 report by the Committee on Evaluation of Sound Spectrograms, it could only be considered, at best, an "empirical art" (p. 1) and not a "fully developed technology based on science" (p. 11).

Replacing Kersta with Statistics

For voiceprinting to evolve from craft to method, Kersta had to be removed from the equation. In the continued absence of a scientific consensus around the measurable relations between voice and identity, however, the only way forward was to mechanize the process of comparison itself. It did not suffice that voice was rendered a mechanistic object to be analyzed. Its interpretation also had to be delegated to machinic processes and thus dissociated with the subjectivity of Kersta. After all, the "ultimate outsider is the machine" (Porter, 1995, p. 85).

The solution that eventually emerged at the turn of the millennium was automatic speaker recognition.[8] The National Institute of Standards and Technology (NIST), for instance, announced the first Speaker Recognition Evaluation (SRE) to test early ASR models in 1996.[9] Like voiceprinting, early ASR was still concerned with identifying speakers through an analysis of their vocal features. But the key development from Kersta's voiceprint technique was the introduction of automated statistical analyses (pattern recognition models) to standardize the process of analysis and thus eliminate errors that would inevitably emerge in subjective human interpretations. By thus framing the task of speaker identification as a simpler pattern recognition problem, the ASR community of scientists and engineers sought to replace the contestable and precarious "expertise" of Kersta with the mechanical rationality of statistical models. Both Kersta and his training program would no longer be necessary. Under the banner of ASR, voiceprinting could be reclaimed as mechanically objective through numbers, statistics, and software: Interpretation would be adequately mechanized.

As Porter (1995) argues, however, mechanical objectivity is a response to a lack of trust. It does not eradicate subjectivity but simply manages it. Indeed, rules are not channels for accessing truth but rather technologies of convenience and standardization. As such, their primary goal is to make the contexts to which they are applied *manageable*. This was indeed the case for the ASR community. The introduction of statistics and pattern recognition algorithms not only replaced the contestable expertise of Kersta but also diminished the demand for a rigorously established consensus among relevant scientific communities. Through the lens of statistics, "scientific knowledge depended on a correct approach, and this meant, first of all, the taming of human subjectivity" (p. 21). The new focus was thus on standardizing the *process* of interpretation and less on testing its validity.

In this way, the shift to statistical frames allowed ASR to "resolve" two fundamental issues central to the original failure of the voiceprint technique:

1. It replaced the "expertise" of Kersta with a formalized statistical approach.
2. It circumvented the need for scientific consensus by reframing it as a statistical pattern recognition problem.

The second achievement was arguably even more instrumental than the first. Recall the FBI-commissioned report from 1979: The first and most significant recommendation by the Committee on Evaluation of Sound Spectrograms for legitimizing the voiceprint technique was the "Scientific Understanding of Voice Identification" (p. 63). Although there was a general understanding that *some* relations undeniably existed between vocal tracts and the vocal sounds produced through them, there was no concrete research showing what exactly those relations were or how to reliably measure them. Although the committee acknowledged that statistics could serve as a temporary fix in this absence of theory, it emphasized that a basic scientific understanding of how voices relate to bodies and identities would be necessary for establishing its legitimacy.

The value of ASR was that it enabled its developers and users to "stay close" to this scientific hypothesis without actually proving or even testing it. Indeed, a term that was used closely with ASR was "voice biometrics." Positioning particular vocal features as stable bodily signals (like fingerprints and iris patterns), the ASR community invoked the hypothesis that the distinctness of voice signals stem from unique vocal anatomies without ever directly explaining the physiological relations between body and voice. Instead, the approach was to *retroactively* interpret the statistical correlations produced through ASR applications as evidence for this hypothesis. In this way, statistical "reliability"—the reproducibility of measurement—would replace Kersta's expertise. And statistical "validity"—the degree to which an instrument measures what it is intended to measure—would be claimed through the *presumed* relations between voice and body. Eliding the committee's first and most foundational recommendation, speaker identification would thus be further entrenched as an art to be mastered and a statistical problem to be solved, but no longer a scientific hypothesis to be tested.

The Rise of ASR and New Experts

As a result of these developments, the disavowal of voiceprinting in the 1970s was somewhat overturned by the early 2000s, especially in non-testimonial settings. ASR technologies were used extensively, for instance, as parts of counterterrorism operations by the US National

Security Agency in 2004 (Kofman, 2018a). In 2014, Interpol established the Speaker Identification Integrated Project, which "allow[ed] uploads and downloads of samples from 192 law enforcement agencies across the world" (Kofman, 2018b). Securus Technologies, one of the largest prison telecommunication firms in the US, released Investigator Pro, a voice biometric search engine to be used on prison phone lines, in 2018 (Securus Technologies, 2018). In 2020, Nuance Communications, the world's leading voice biometrics company, claimed to have collected more than six hundred million voiceprints for its voice biometric database (Nuance Communications, 2020). The UK similarly amassed millions of its own citizens' voiceprints through the integration of voice biometric systems into government services such as HM Revenues and Customs phone lines (Peachey, 2019). Similar reports of large-scale voiceprint collections were made by China and Turkey regarding their mobile phone networks (Kofman, 2018a, 2018b).

The newfound reproducibility of mechanized interpretation coupled with the reframing of speaker recognition as a statistical challenge seemed to have effectively resuscitated the use of forensic speaker identification technologies around the world, just under a different banner. Indeed, as Antonio Moreno, vice president of Agnitio,[10] the company behind the world's most widely used ASR system, BATVOX, declared, "Our system . . . is measurable and can be reproduced: two different operators will get the same result from the system" (Catanzaro et al., 2017). But as Porter (1995) asserts, "Mechanical objectivity can never be purely mechanical" (p. 5). Even BATVOX, which is used by Interpol and is "present in more than 50% of forensic labs worldwide . . . [with] more than 100 installations internationally and used in court over 35 countries" ("Nuance Public Safety and Security Forensic Biometric Solutions," n.d.), is contingent on its operators receiving extensive training through the BAT School Certification Program.

According to the brochure published by Agnitio, the certification program is broken down into three levels of user proficiency—basic, advanced, and expert (see table 1.1)—and is intended to be completed in "3 years provided that the user has intensive contact with BATVOX" ("BATVOX Advanced School," 2012, p. 3). It also provides "laboratories and official organizations to qualify, in an objective and independent way, their level

Table 1.1
BATVOX certification levels

Certification Levels	Proficiency
User Level 1 (Basic)	The user is prepared to perform experiments with BATVOX, prepare voice corpus, and evaluate the system for standard or noncomplex conditions
User Level 2 (Advanced)	The user is qualified to use BATVOX as a support to forensic real-life investigation activities (for law enforcement and intelligence groups) and to design and perform experiments for unknown new conditions
User Level 3 (Expert)	The user is ready to prepare forensic grade results of automatic speaker recognition using BATVOX. Can be used by certified eyewitness experts in conjunction with other technologies to present evidence in court.

Adapted directly from page 3 of Agnitio's "BATVOX Advanced School" (2012) brochure.

of knowledge in the use of BATVOX" (p. 1). The company, however, also hedges its claims:

> It is important to remark that BAT School Certification program does not qualify to present audio evidence in court. This can only be done if the user is entitled (according to local regulations) to provide expert evidence in legal courts. Certification Level 3 guarantees that the user can use BATVOX correctly. However, the presentation of Speaker ID evidence in court might require additional expertise and training . . . AGNITIO discourages the use of BATVOX in court by non-forensic experts, even if the user has gone through this course, and reserves the right to cancel or not renew subscription if the system is used improperly. (p. 3)

In other words, Agnitio is selling another *type* of expertise. The intensive three-year program suggests that BATVOX is anything but a straightforward system that operates according to clear and formalized rules. Rather, it is presented as another complex tool to be learned and mastered in the same way that Kersta's students had to study his methods over the course of two years (or longer). As the committee commented in 1979 of Kersta's voiceprinting technique, "So much experience would not be needed if explicit and objective criteria were available" (p. 14). The same can be said for BATVOX. Indeed, there is an inherent tension between mechanical objectivity and extensive training requirements. It is thus more accurate to understand Agnitio and BATVOX as simply replacing the role of Kersta rather than addressing the underlying limitations of his method.

The program also clearly differentiates "technical" expertise in BATVOX from "disciplinary" expertise, the latter of which Agnitio emphasizes is crucial for providing evidence in court. Although Agnitio is highly incentivized to sell its certification program, it is also important for the company to acknowledge its limitations, given the high-stakes settings for which BATVOX is meant to be used. Indeed, the standardization of method through rigorous training is never a substitute for a lack of scientific understanding. In the same way an expert lab technician is not equivalent to a seasoned chemist, technical proficiency in BATVOX says little about the operator's command of the scientific domain to which the technology is applied. The situation gets even more complicated when considering that it was reported in 2017 that only 20–25 percent of BATVOX users had completed the certification program (Catanzaro et al., 2017). Simply put, most users at the time did not even acquire a basic level of *technical* know-how. Taken together with the fact that BATVOX is used by more than 50 percent of the world's forensic labs and is present across thirty-five different countries ("Nuance Public Safety and Security Forensic Biometric Solutions," n.d.), this raises serious concerns about how these systems are made available to users who have insufficient levels of technical expertise.

Unfortunately, this is a common symptom of automated systems. Although they are sold on the promise of efficiency and accessibility, they often require new forms of labor and expertise to operate "correctly"—a requirement that is frequently elided based on the misguided belief that automated systems produce more accurate results (e.g., automation bias).[11] In this way, the supplanting of subjectivity with supposedly more objective (consistent) techniques often only amounts to replacing one form of expertise with the requirement for another.

As such, although ASR mechanized the process of comparing different voice samples through recording devices, speech processing software, and statistical models, it never fully escaped the necessity and grasp of human experts. It distanced speaker identification from Kersta and his spectrographic methods but established new experts and standards in the process. As speech signal processing researchers John H. L. Hansen and Taufiq Hasan (2015) write, "Today, forensic speaker identification is commonly performed by expert phoneticians who generally have background in linguistics and statistics. This is a very complex procedure, and varies among practitioners" (p. 80). Indeed, as ASR developed, it was ultimately the statisticians—and, with the development of deep learning techniques, machine learning practitioners—who took the reins as the new experts of speaker recognition. So long as the *belief* that voices carry signals of unique bodies was intact, understanding the physiological production of voices was no longer relevant to the underlying task. The voice could be quantified through its acoustic features and then treated purely as a numerical site of statistical analysis.

Mechanical Solutions

ASR is now generally considered an umbrella term for two "distinct" technical tasks: speaker *verification* and speaker *identification*.[12] Verification, also often referred to as "voice ID," is generally used as a security feature for access to personal accounts (e.g., banking) and digital services (e.g., voice assistants such as Siri and Alexa). Here, an individual consents to and participates in the construction of a "voice template," in which several acoustic features are extracted from a voice sample during an initial *enrollment* phase. This template serves as the stable reference against which future samples or "utterances" are compared during the subsequent *verification* phase. In other words, the task becomes a relatively simple one-to-one (1:1) matching scenario: Does the new sample match the prerecorded template?

Although verification is primarily used as a supplementary security feature in multifactor authentication systems, speaker identification is considered more widely applicable. Here, the scenario is one in which there is an unknown speaker in an audio sample, and the task is to determine whether this unknown speaker matches a known speaker from a set of pre-collected templates (1:N). If the unknown speaker is known to be within the reference set, it is considered "closed-set speaker identification"; if there is the possibility that the unknown speaker is not included, it is referred to as "open-set identification" (Faundez-Zanuy & Monte-Moreno, 2005, p. 2). Both verification and identification can be text dependent or text independent. As the names suggest, text dependent requires the text spoken during the enrollment and authentication phases to be identical, whereas text independent does not. As such, text-independent recognition relies solely on comparing the extracted fundamental acoustic features.

Even before the application of deep learning to speaker recognition, text-dependent verification under controlled settings was largely considered a "solvable" engineering problem.[13] It effectively bounded speaker recognition within a manageable setting in which the primary task was to compare one utterance with another and compute the similarities of their underlying acoustic features. In this way, the voice could be rendered a *signal* in the purest sense. It was no longer attached to the intricacies of theoretical inquiry: What are the relations between body, voice, and identity? Explanations became unnecessary; voice just needed to be bounded tightly enough to fit the parameters and goal of the task: Does $x_{\text{template}} = y_{\text{template}}$?

Identification, however, was considered more complex. The task was a problem not only of pattern matching but also of comparing *similarities* and *typicalities*. As forensics researcher Geoffrey Stewart Morrison and criminology professor William C. Thompson (2017) write, two questions were central to speaker identification:

1. What is the probability of obtaining the observed properties of voice on the questioned-speaker recording if it were produced by the known speaker?
2. What is the probability of obtaining the observed properties of the voice on the questioned-speaker recording if it were produced not by the known speaker but by some other speaker from the relevant population? (p. 349)

The first question sought to quantify similarity, whereas the second sought to quantify typicality. In statistics, the relationship between these two numbers is referred to as the "likelihood ratio" (p. 348), and this was used as the primary quantitative measure for evaluating the strength of evidence in automatic forensic speaker identification.

Consider the following simplified example: There is a speech sample in question of a man speaking in Korean. The setting in which this speech sample was known to be uttered is Seoul, South Korea, where most people communicate in Korean. A statistical analysis shows that the vocal features extracted from the speech sample also happen to be highly typical of those of men speaking Korean in Seoul. Given this typicality and the large sample size, this would make the probability for the second question very high. As such, even if the similarity of the speech sample to the template of a potential suspect was high, the high probability for typicality would offset the strength of the evidence. In contrast, if the same speech sample was known to be uttered in the West Village of Manhattan, and a statistical analysis shows that the vocal features extracted from the speech sample are relatively atypical of those of the comparatively fewer men speaking Korean in the West Village, this would make the probability for the second question quite low. As such, if the similarity of the speech sample to the template of a potential suspect was high, the evidence would also be considered very strong.

In this way, identification not only required the distillation of voice into mechanical templates to be compared but also involved a *calculus of context*. Indeed, a significant part of the identification problem before deep learning was concerned with controlling and quantifying "backgrounds" and related media of all sorts: technological channels, duration of audio, linguistic typicality, relevant speaker populations, cultural contexts, psychological mediators, social settings, and so forth (e.g., Hansen & Hasan, 2015; Morrison & Thompson, 2017; Reynolds et al., 2000). Identification research thus centered approaches for reducing contextual noise and constructing ideal statistical settings. The objective was to reframe the identification problem into one that could more closely resemble the controlled structure of the verification task.

The primary distinction between verification and identification thus boiled down to their varying manageabilities. Whereas verification could be neatly bounded into a task within a controlled setting, identification could not. The underlying question for both was how to make the voice *useful* (as security feature or forensic evidence), which actually required moving further away from *understanding* the voice. Indeed, from the specific standpoint of the tasks at hand, scientific inquiry appeared unproductive. The objective was to flatten voice and the relevant contexts around it into numbers to make them amenable to statistical frames. Here, theories of voice became gratuitous—utility and manageability became the primary parameters of relevance.

Deep Learning: From Control to Scale

The development of *deep learning* techniques—a subset of machine learning that uses artificial neural networks with multiple layers to analyze vast amounts of data—further solidified the fungibility of theory while also resolving some of the complications around earlier ASR approaches. The most salient difference between the previous machine learning techniques of ASR (e.g., Gaussian Mixture Models) and the newer deep learning applications for speaker recognition revolved around a practice called "feature engineering." In the simplest terms, this refers to a preprocessing step in which data is transformed into a more legible set of inputs composed of relevant features. Whereas feature engineering is a manual process in other ML approaches, the introduction of "representation learning" in deep learning applications turned it into an automated process. As such, in the pre–deep learning context of ASR, manual feature engineering still required ML practitioners to somewhat "engage" with the voice data. Although the goal was ultimately to distill voice recordings into numerical formats that represented its most relevant acoustic features, ML practitioners still needed to control and decide which parts of the voice would be made useful.

A primary shortcoming of identification tasks in ASR before deep learning was thus that their performance would "degrade rapidly in real world noise" (Nagrani et al., 2017, p. 1). Even the success of verification relied on the task being structured within bounded settings: Both the initial enrollment and subsequent authentication of speakers were conducted in controlled (i.e., non-noisy or consistently noisy) environments. Pre–deep learning verification would not have been considered a feasible task if these other parameters could not be controlled for. Both the feasibility of identification and verification were thus inextricably tied to the manageability of voice and its environments. They were unrealistic tasks outside of controlled settings with clean recordings and savvy feature engineering.

Deep learning effectively eliminated these steps of decision-making. In fact, the objective changed from tightly controlling the environment to intervening as little as possible. The primary action shifted from *control* to *collect*. As computer scientists Arsha Nagrani, Joon Son Chung, and Andrew Zisserman (2017) write, "One of the most important ingredients for the success of [deep learning] . . . is the availability of large training datasets" (p. 1). Indeed, deep learning transformed the crux of the problem from statistical management to data collection. The goal shifted from tightly constraining the terms of the recognition tasks through various processes of cleaning, structuring, and distillation to intentionally selecting for and collecting as much "messy" and so-called "real-world" data as possible. The underlying sentiment was that data should "speak for itself."

The previously mentioned NIST SRE challenges provided some of these early large-scale datasets, but they were not made available to the broader research community. The "Speakers in the Wild" (SITW) dataset, which "consists of recordings of 299 speakers, with an average of eight different sessions per person" across "unconstrained or 'wild' conditions" (McLaren et al., 2016, p. 818), is thus generally considered the first *publicly* available speaker recognition dataset. SITW, however, was limited in scope: It was hand annotated, which was not considered a scalable approach for constructing the larger scale datasets required for deep learning.

VoxCeleb, widely considered the first publicly available *large-scale* speaker recognition machine learning dataset, addressed this limitation. Using facial recognition models to automate the process of annotating speakers, the creators of VoxCeleb were able to construct a significantly larger dataset containing more than a hundred thousand utterances across 1,251 celebrities extracted from videos uploaded to YouTube. The videos were neither controlled for audio clarity nor processed for consistency post collection. Described by its creators as "a large scale text-independent speaker identification dataset collected 'in the wild'" (Nagrani et al., 2017, p. 1), VoxCeleb became central to setting the terms for what speaker recognition would look like in the context of deep learning.[14]

Importantly, the previous distinctions between verification and identification also became less clear. What differentiated the feasibility of each subtask before deep learning was that verification could be designed into a manageable structure (and therefore considered a feasible engineering problem), whereas identification could not. Previously, the distinction between verification and identification was 1:1 versus 1:N *only in concept.* In practice, the distinction really came down to their varying manageabilities, which ultimately constructed each as separate tasks with vastly varying structures. In the context of deep learning, however, both verification and identification came to rely on models that draw from the same underlying "noisy" dataset: The difference *truly* became a matter of 1:1 versus 1:N. Indeed, as computer scientists Zhongxin Bai and Xiao-Lei Zhang (2021) write in their overview of speaker recognition based on deep learning, "Speaker verification is a special case of the open set speaker identification problem with $K=1$" (p. 3), where K denotes the number of enrollment speakers.

In this way, the complications surrounding speaker identification were addressed through an astonishingly simple "brute force"[15] approach. As previously un-processable amounts of data became not only usable but also *useful* in the context of deep learning, the previous distinctions between verification and identification also became somewhat trivial. The cumbersome calculus of contexts was no longer necessary for identification—it simply became the broader domain of recognition that includes the verification task. Once considered problematic, "noisy" data was now seen as containing "valuable variations" that deep neural networks could parse and exploit. The key to identification was not a matter of *control*, as previously thought, but rather a problem of *scale*.

Deep learning thus fundamentally changed the terms of automatic speaker recognition.[16] The transformation of the task from an interventionist engineering problem of control to an arguably simpler data collection problem of scale effectively rendered any remaining requirement of domain understanding and/or scientific theory superfluous. Even the nature of "expertise" changed. It was further displaced from the spheres of voice, body, sound, and even acoustics, and reclaimed in the context of computer science. It is telling that all of the creators of VoxCeleb are computer scientists, some of whom are better known for their contributions to computer vision research than speaker recognition. This makes sense in the context of deep learning: The point is to get as far away from the materialities of voice as possible and instead to treat the voice as just another site of data collection amenable to algorithmic analysis.

Philosopher of science Mel Andrews (2024) critiques this "theory-free ideal," which has become part and parcel of many machine learning communities well beyond speaker recognition, as fundamentally different from even traditional statistics. Here, they describe deep learning as exhibiting "a novel kind of epistemic opacity to its deployers" (p. 7). Indeed, as philosophers of science Sanja Srećković, Andrea Berber, and Nenad Filipović (2021) also write, "For traditional statistics, standard models rely on the representation of underlying causal mechanisms, and they are used for retrospective testing of an already existing set of causal hypotheses. On the other hand, ML models are constructed based on data instead of theoretical assumptions about the target system" (p. 166). Here, explanation is not only *deprioritized* in favor of utility but also *impossible*, given the inherently uninterpretable and abstract nature of the algorithmic processes of deep learning. Action—the ability to use and act upon voice data—becomes the only possible goal.

* * *

The evolution of speaker recognition technologies, from the rudimentary sound writers of the nineteenth century to mid-twentieth-century sound spectrographs and ultimately the statistical machine learning approaches post 2000, follows a trajectory of how voice is fitted to new frames for the purposes of managing it. To be sure, there are numerous other histories through which the voice was similarly examined through various mediums to serve different purposes. But the unfolding story of speaker recognition technologies, which I roughly bookend with the voicegram, on one end, and VoxCeleb, on the other, is particularly useful to contextualize the current state of the voice AI industry. Indeed, this history engages resilient folk theories that associate voice, body, and identity, without ever concretely establishing their relations, as well as attempts to enact mechanistic frames of voice that ultimately push us away from its material qualities, all of which remain central to the underlying logics of voice AI today. As I show in this chapter, "frames of voice" are always in service of managing it and everything else it can be made to stand in for.

As we will read in parts II and III, all attempts to interface voice with artificial intelligence will also always be projects of reframing voice as a manageable mechanism to be traced and replicated. Here, notions of "measurability," "replicability," and even "progress-ability" are crucial features through which the voice becomes not only a machine *learnable* modality in contemporary artificial intelligence systems but also a concept that fits into the capitalistic paradigms of "growth" and "progress." As will be evident, the broader imaginaries upon which speaker recognition technologies depend, the conditions that enabled their development, and the techniques through which they have been rationalized comprise an important backdrop for the stories to unfold.

Interlude 1 Machine Voicescapes

"To listen to listening" is to study the world in and through which the listener listens. In the field of sound studies, this orientation has served to shift attention away from the notion of a sonic truth and toward *ways of listening*, centering listeners and the worlds they inhabit as the source of sonic knowledge (e.g., Eidsheim, 2018; Stoever, 2017). Indeed, sounds only exist and are made meaningful in the worlds of listeners.

I am reminded here of the term "soundscape," a term that is often used to describe an acoustic environment—the sonic counterpart of "landscape," if you will. R. Murray Schafer (1977/1993) first popularized this term through his work *The Soundscape: Our Sonic Environment and the Tuning of the World*, in which he examines the effects of urbanization on the world's sounds. Distinguishing "hi-fi" soundscapes—environments that have a positive and high signal-to-noise ratio, which means that discrete sounds can be heard clearly due to the lack of background noise (think of the sound of birds in a quiet meadow)—from "lo-fi" ones—settings with a low or negative signal-to-noise ratio, in which signals are obscured by too many sounds (think of trying to have a conversation in a busy convention center)—Schafer was responding to a particular moment in history characterized by rapid urbanization. The increasing sounds of industrial growth and urban sprawl became an indispensable part of people's everyday sonic worlds. For Schafer, soundscapes provided a frame to theorize the changing sonic topography around him.

More recently, however, the term has been adopted across various disciplines such as computer science and music production to refer more broadly to an immersive-feeling acoustic environment. In this way, the term "soundscape" lost some of its original specificity and the underlying commentary baked into its use: Schafer's disdain for the encroaching sounds of urbanization. Indeed, Schafer's use of the term fundamentally centered his own subjective embodied experience of the acoustic environment to/in which he listened. Although the specificity of Schafer's aversion to the sounds of the city was lost in the term's eventual adoption in other fields, the sense of a perceived immersion in sound and the centrality of the embodied listening subject in this perception still remain vital to its conceptual framework. Soundscape describes more than just an acoustic environment. It is the sonic knowledge and experience born out of the *relationship* between listeners and their worlds.

The following two chapters are precisely about these relationships—not only between listening humans and the soundscapes that mediate their ways of listening but also between "listening machines" and the soundscapes that humans construct for them. Here, I suggest the term "voicescape" to center listeners and the broader ecology of social, cultural, biological, and technological conditions through which voices are made meaningful to them (Ceraso, 2022). Voices, too, only exist in the worlds of listeners.

A *machine* voicescape, then, centers the machine and the ways in which voices are made stable, legible, and audible to it. If chapter 1 recounted a history of how different frames of voice were assembled and discarded for managing voice in the context of speaker recognition technologies, the chapters in part II center how those frames are made useful to machines designed to listen to our voices in the context of the voice AI industry. Together, they chart the story of the "listening" machine, the worlds—that is, truths—that are constructed around it, and the actors that collectively define, determine, and police the borders of these machine worlds. Here, we will examine both the processes of voicescap*ing* for machines and the resultant voicescapes that emerge.

This examination will thus inevitably also necessitate investigating how machine "truths" are created, where they are stored, how they are maintained, and who gets to manage them. This is the topic of chapter 2. Indeed, machine truths exist on their own terms. And we will need to understand what these terms are to also chart the ways that machine listening is positioned as a means of accessing (constructing) those truths. In chapter 3, we will build on this understanding to examine how such truths are constructed for machines designed to "hear" emotions in our voices.

II Listening Machines

2 Machines Learning to Listen

Machine Worlds

In one of the earliest attempts to understand what is now grouped under the banner of machine learning, Oliver Selfridge, a pattern recognition researcher, and Ulric Neisser, a cognitive psychologist, wrote that "any system must fail if it tries to specify every detail of a procedure for identifying patterns that are themselves defined only ostensively. A pattern-recognition system must learn. But how much?" (1960, p. 65). Indeed, this question of "how" appropriately directs attention to the epistemological foundations of machine learning. Arguably more vital, however, is the ontological question of "what."

What is a machine being made to learn? For Selfridge and Neisser, as well as the broader community of early pattern recognition researchers in the late 1950s and early 1960s, the answer was optical character recognition (OCR): the automatic recognition of hand printed letters via computational statistics.[1] Despite the challenges that drove this scientific community at the time, OCR presented a relatively easy "what." It was easy in the sense that written letters of the alphabet—a bounded sequence of twenty-six distinctly shaped characters that can be described or deduced with descriptive features such as whether the letter has a vertical line, crossbar, or "a concavity at the top" (p. 66)[2]—comprised a stable reference for constructing a "ground truth" for OCR.

Ground truth has now become a familiar concept in the machine learning community, but it's a term that originates in the fields of meteorology and remote sensing. There, it refers to information that is collected directly on site so that image data can be linked to real features and materials *literally* on the ground. It is, in this way, a referential concept that not only describes the relationship between data and a real-world phenomenon or object that the data is meant to represent—for instance, a type of soil that exists in a particular location—but also *verifies* it. Indeed, ground truth in meteorology is the result of a process of *corroboration,* in which data are subsequently substantiated through on-site observations and measurements.

The adoption of ground truth in machine learning flips this sequence of truth making. Here, ground truth refers to "the dataset used to [both] *train* and *validate* [machine learning] models" (Sadowski, 2025, p. 5, emphasis added). As such, it denotes not only a corroborated result (validation) but also a reference that the model is *calibrated toward* (training). It predefines the "assumed truth" and possibilities a machine learning system is designed to reproduce in which, upon successful reproduction, the "assumed truth" becomes simply "truth"—at least in the context of the problem space to which the machine learning system is applied. In the example of the machines designed to recognize handwritten letters in the 1960s, then, the English alphabet would comprise the referential boundaries, and a dataset containing images of handwritten letters of the English alphabet would comprise the material of the ground truth. ML ground truths can thus be seen as a collection of a priori conditions, assumptions, and materializations that determine what the machine learning model can/will "know" (e.g., images of handwritten letters of the English alphabet for 1960s OCR but not of *hangul,* the Korean alphabet). It governs the boundaries of machine worlds.

Learnability and Ground Truth Tracings

In this chapter, I examine how such ground truths are constructed in the context of machine *listening* voice AI systems. If the previous chapter charted a history of how voices were rendered mechanically legible/audible in the context of speaker recognition technologies, here, I shift my attention to the *process* of mechanically "listening to voices" and how that process is transformed into a machine *learnable* task.

It is important to note that I differentiate my use of "learnability" from the more technical use of the term invoked in traditional machine learning practice such as "probably approximately correct" (PAC) learnability. PAC learnability describes a specific statistical situation in which a hypothesis is considered learnable when there is a relatively high probability that the "learning function" will have a low generalization error (Shalev-Shwartz & Ben-David, 2014, p. 22). My use of "learnability" refers more literally to the possibility of formulating a machine learning problem—that is, the degree to which a real-world problem can be conceptualized, translated, and formalized into the operational framework of machine learning as ground truth. Evidently, this framing of learnability does not require a direct mathematical engagement with ML models and algorithms. Rather, it takes a *sociotechnical* view that focuses on the human practices of conceptualization and translation that occur in reconceptualizing qualitative phenomena into quantitative machine learning frameworks.

Building on this broader definition, I introduce one approach for qualitatively tracing these "learning"—that is, ground-truthing—processes, which I refer to simply as "ground truth tracings." This approach is an outcome of my fieldwork in the voice AI industry between October 2021 and June 2022. During this time, I spent hundreds of hours speaking with and interviewing individuals involved in various parts of the development pipeline for machine listening voice AI systems. These individuals occupied various positions, spanning founders, CEOs, senior vice presidents, machine learning engineers, computer scientists, data scientists, product developers, project managers, business development managers, database managers, brand evangelists, staff developers, and venture capital investors. The companies that these industry professionals work(ed) for range(d) from well-known big tech or VC firms to smaller but still influential industry leaders in voice AI technologies, as well as a few emerging startups. I reached out to some of these individuals through their LinkedIn pages and Twitter (now X) accounts. Others I met by participating in flagship industry conferences such as the VOICE Summit and VOICE Global, attending webinars organized by various voice AI companies, and joining invited talks where I learned about existing state-of-the-art (SOTA) systems and future visions for the industry.

The most important insight I took away from this fieldwork was that conceptions and expectations around what "listening" and "voice" mean in the context of machine learning, or more specifically machine listening voice AI systems, vary significantly based on who one talks to and the contexts in which they are conceived. For instance, there were product developers and business development managers who were excited about voice

AI systems that could purportedly predict the emotional state of a speaker in real time, estimate the likelihood of a borrower defaulting on a loan, match romantic partners, and (as we will discuss in this chapter) forecast how well a prospective candidate would fit in with a certain team all by simply processing voice recordings using ML models. But there were also engineers and research scientists who represented groups that expressed a critical skepticism toward the current capabilities of not just machine listening voice AI technologies but also the current capabilities of machine learning, writ large.

These tensions are inevitable outgrowths of a technical community that depends on the distillation of complex concepts—such as voice or, more specifically, "vocal audibility"—into machine legible formats (data). Indeed, unlike in the OCR community of the 1960s, in which the alphabet presented a relatively straightforward reference from which a ground truth could be constructed, the voice and how we listen to it is notoriously diverse and complex.[3] Numerous scholars have pointed to the voice as a fluctuating phenomenon that is both produced and heard differently according to the sociocultural (e.g., Eidsheim, 2018; Semel, 2022; Stoever, 2017), biological (e.g., Kreiman & Sidtis, 2011), and physical (e.g., Weidman, 2015) conditions in which it is voiced and listened to. As such, neither voices nor our listening practices comprise reliable sources from which to derive stable references upon which a machine can "learn." This makes the voice AI industry an especially challenging but also compelling space to examine the tensions of ground-truthing practices in machine learning.

This complexity is further compounded by the fact that ground truths are ultimately reflections of the people who design them and the tasks for which they are designed. As Passi and Sengers (2020) observe in a different machine learning community, the definition of a "functioning chatbot" can differ widely within the context of a single business, depending on whether one asks the "data science" team or the "business strategy" team. These material differences across individuals, which can pertain to opinions around the technical feasibility of a task, scope of work, ethical boundaries, resource efficiencies, and domain applicability, among others, drastically change how one views the concept that is to be machine learned (Orr & Crawford, 2024). In this way, both the ontological complexity of a concept as well as the practical constraints of development collectively determine the borders of machine learnability.

As a method, GTT attends closely to these boundaries. It centers the material and situated practices of human decision-making, positioning machine learning systems and the specific tasks they are meant to address as inherently bounded within the local contexts in/for which they are designed. In this way, although the specific context of voice AI presents a compelling case study for GTT, the approach can also inform evaluations of machine learning systems more broadly. As technology researcher Jathan Sadowski (2025) writes, "There is no escaping the necessity of ground truths that have been created and certified by people. Even if they are not used to train the model, they must be used to evaluate and adjust the model. The central reference point is always how close the model gets to reproducing the chosen ground truth in its outputs" (p. 6).[4] Examining machine learning models through the aperture of ground truths thus allows for a productive distancing from the narrower technical discourse of model performance and algorithmic efficiency, instead foregrounding the broader socio-material practices of data collection, preparation, and maintenance while still engaging with the logical step-by-step sequences related to data analysis that must occur for a machine learning system to function. And as a result, it provides an opportunity for interdisciplinary dialogue between engineers/developers and social scientists/humanists to collectively think through the broader sociotechnical assemblage of humans, machines, imaginaries, and infrastructures that "learnability" is contingent upon.

These collaborative efforts become especially important as machine learning is implemented for use cases with increasingly higher stakes. In his 1963 paper "The Imitation of Man by Machine," Neisser emphasized that the fundamentally different underlying processes by which humans learn and machines are programmed to "learn" have serious consequences in real-world applications. We must not make light of these differences. Although variations in human and machine learning may be relatively trivial for the purposes of straightforward *technical* pattern recognition tasks (e.g., OCR), they become, as Neisser asserts, tremendously apparent and important when computers take on the role of adjudicating *social* decisions—that is, decisions without clear answers mediated by power with people at stake. The voice AI technologies examined in this book are representative of these kinds of systems. To adequately understand and evaluate their impact, it is thus imperative that we examine how the voicescapes through which they process and "listen" to the worlds around them are constructed as ground truth.

Formulating Machine Learnable Problems

In the words of sociologist Adrian Mackenzie (2017), "To understand what machines can learn, we need to look at how they have been drawn, designed, or formalized" (p. 49). This can be done by taking a closer look at the *problem* around which a machine learning system is developed. Science and technology scholar Florian Jaton (2021) writes that a "problem an algorithm is designed to solve does not preexist: it has to be produced during what one may call a 'problematization process'—a succession of collective practices that aim to empirically define the *terms* of a problem to be solved" (p. 82). A central step in this problematization process is ground-truthing, which, as mentioned earlier, refers to the process of building a referential repository (ground truth) that serves as the base from which machine learning algorithms are derived. In this way, ground truth is quite literally where the truth and possibilities of an algorithmic model are grounded. Most importantly, this means that a problem of interest and the method of addressing that problem must first be defined and determined through data in the ground truth. Ground-truthing thus constitutes a task-bounding process and a form of intentional biasing that hardlines the limits of the algorithm and the possible range of outcomes for a machine learning system. Ground truths establish the contours of the problems to be solved.

This specificity of machine learning is sometimes referred to in industry discourse as "artificial narrow intelligence" (ANI). The modifier "narrow" inserted between the more common phrase "artificial intelligence" is meant to emphasize the "goal-oriented version of AI designed to better perform a single task" (Kanade, 2022). What is misleading about this rhetorical distinction from other forms of AI such as "artificial general intelligence," however, is that it attempts to expand the domain of artificial intelligence through the implication of another more adaptive version of AI that goes beyond the capabilities of ANI. To be clear, there is no such version yet.[5] AGI, which refers broadly to a *vision* of AI that has the capacity to emulate human cognition and adapt to *any* complex problem, is still a loosely defined concept and an unrealized ambition within the machine learning community. In fact, as many critical AI researchers have pointed out, the hype around AGI (most notably driven by companies such as OpenAI) distracts from the actual potentials and harms of existing AI/ML applications and the broader ethical, social, and practical questions they pose (Johnson, 2022). It is thus important to emphasize that in the current state of AI/ML research, ANI comprises the *entirety* of AI applications (Chollet, 2019). I accentuate this point here because it illuminates the centrality of ground-truthing practices in how any current AI/ML system is made to execute a task. As Jaton (2021) writes, "*We get the algorithms of our ground truths*" (p. 81).[6]

Indeed, examining an ML system through the aperture of ground truths allows one to examine the practical steps by which a claim made around a certain technology's capabilities is "ground-truth*ed*." Here, the underlying questions, in their simplest form, are: (1) Is this a learnable problem? (2) If so, how? (3) If not, why not? In the same way that ground truths must first be formed as a base upon which ML algorithms can be developed, qualitative tracings of ground truths (GTT) can also retroactively be *per*formed to interrogate the logical sequences and materials through which a ground truth is constructed for that claim.[7]

Example: Visual Saliency

Momentarily departing from the world of sound, Jaton's (2021) discussion of visual saliency detection presents a useful example to contextualize this frame. He writes that in contrast to high-level detection algorithms such as those for face detection, which are task specific and thus conceptually easier to construct ground truths for, low-level saliency detection—detecting which parts of an image are "salient"—has "no 'natural' ground truth allowing the design and evaluation of computational models" (p. 56). This impossibility lies in the obvious fact that "what is considered as salient in a natural image tends to change from person to person" (p. 57) and, of course, situation to situation. Performing GTT is thus a fundamental task for ML engineers because it allows them to either dismiss a problem as one in which a ground truth cannot possibly be established—that is, an "unlearnable" problem—or find a way to redefine the problem in such a way that a ground truth can be established.

Liu et al. (2011) tackled the problem of visual saliency detection by reframing the task from simply recognizing salient parts of an image, which was deemed to have too much statistical variance, to detecting "*the most salient object* within a given digital image" (Jaton, 2021, p. 57). This still

Figure 2.1
Left: An example of an image in which there is an obvious distinction between figure and ground. Right: An example of an image in which the distinction between figure and ground is not obvious. Liu et al. (2011) developed a ground truth dataset that was composed of images like that on the left and not that on the right.

has the potential for disagreement between labelers, but it can decrease the variance for images in which there is a relatively obvious distinction of figure versus ground (figure 2.1).

By thus repositioning saliency within this binary framework, it became possible to conceive of and build a ground truth dataset. Of course, this is not to say that in the process of developing a ground truth, negotiations, disagreements, or other complexities related to labeling and classification do not occur. The establishment of a ground truth should *never* be confused with a "true" quantification or datafication of a real-world phenomenon. Rather, it is the translation and flattening of a messy qualitative problem into a usable quantitative reference.[8] As Mulvin (2021) also writes of scientific proxies, ground truths are part of the broader "culturally conditioned practice of consistently using some things to stand in for the world" (p. 5). Indeed, ground-truthing is a process of representation and translation, in which certain fragments of the world are compressed, stabilized, and sanitized to be made legible to/for a machine learning system.

What is implied in ML ground truths is thus not necessarily a representation of "reality" but rather the *translatability* of a problem of interest, which allows it to be legible to and expressed in the language of mathematics.[9] This numericization allows it to become more sharable, comparable, and malleable, thus allowing new statistical instruments to be used in its analysis. The most ideal situation is thus one in which a problem of interest can easily take a quantitative form (i.e., ground-truthed) without losing a significant portion of its qualitative character. The reality, however, is that there are varying degrees to which different qualitative phenomena can be smoothly numericized. Indeed, such translation processes often require reconceptualizing a qualitative problem to fit the quantitative framework through which the problem is formalized (i.e., redefining general "saliency detection" to the binary task of "figure–ground distinction"). As a method, GTT makes this translation process explicit for the purpose of gauging the degree of *ontological dissonance* that is generated in the conversion from "complex entity" to "scriptural form" (Jaton, 2021, p. 229). In this way, it is a mode of interrogating the practical steps that ML practitioners must take in developing usable machine learning datasets and thus "making work" the machine learning systems that are trained on them.

Ontological Dissonance: OCR Versus Facial Criminality

Consider the task of recognizing hand-printed characters that was of interest to the community of early pattern recognition researchers mentioned at the beginning of this chapter. As Selfridge and Neisser (1960) document in "Pattern Recognition by Machine," letters of the alphabet can be described in terms of specific features. An "A," for instance, has a horizontal crossbar near the center and is convex shaped in the upper half and concave shaped in the lower half. Few other hand-printed letters in the alphabet would match this description, and from a regression logic, no other letter would consistently conform. Simply put, the qualitative character of an A can be relatively accurately described in terms of such discrete features, and a dataset that contains many examples of handwritten "As" should be sufficient in representing this pattern. The ontological dissonance between these features, the examples in the dataset, and what we generally understand as an "A" is minimal.

In contrast, consider the controversies around machine learning research that have claimed to be able to predict criminality through a computer vision analysis of facial features (Wu & Zhang, 2016a). Without even beginning to engage the histories of scientific racism in phrenology that such claims invoke, it is possible to examine the technicality of these claims by drawing from GTT: *Is criminality a machine learnable problem?*

Wu and Zhang (2016b) write in their response to the highly publicized critiques of their controversial paper "Automated Inference on Criminality Using Face Images" (2016a) that "taking a court conviction at its face value, i.e., as the 'ground truth' for machine learning, was indeed a serious oversight on our part" (2016b, p. 1). It is interesting to note here that despite the many other refutations made by the authors against their many critics, they do (or perhaps have no other choice but to) accept that a court conviction does not establish an adequate ground truth. This is important because it reflects the efficacy of shared terminology in interdisciplinary dialogue. By couching a sociohistorical critique of court convictions within a machine learning context of technical ground-truthing practices, not only does it give the critique more heft and a technical base from which developers who build machine learning programs can work, but it also allows for a productive rethinking of how feasible the proposed machine learning system is.

Unfortunately, however, despite the authors' admission that using an individual's prior conviction as ground truth for criminality was a "serious oversight," it is telling that they don't (can't) address the development of an alternative ground truth. Instead, the authors double down on their conclusions stating that they "maintain a sober neutrality on whatever [they] might find" (p. 1). This is to be expected, given that the only means through which "criminality" can be recognized is through the very justice system that defines it. There is simply no other way to construct a ground truth for criminality: An individual's criminal status is established only at the moment of conviction. As such, once an individual's prior conviction is accepted as constituting the "criminality" component of the ground truth, a photo of that individual's face is labeled as a "facial instance" of "criminality" to establish the full ground truth of "facial criminality."

GTT foregrounds the examination of these separate human assumptions and practices required to construct ground truths. In this case, it forces an engagement with the underlying *materiality* of "facial criminality" and, in so doing, also provides a concrete angle to challenge it. If it is acknowledged here that "criminality" does not exist outside of the documents and institutions that produce it, then critical attention can be directed explicitly to those documents and institutions. Returning to the original question of whether criminality is a machine learnable problem, the answer is "yes" if we accept these documents to be adequate references, and "no" if we do not. The investigation is no longer an abstract reflection on "facial criminality," but rather a concrete analysis of whether these documents and photos contain measurable signals of criminality that can be parsed using pattern recognition algorithms.

Indeed, we must remember that machine learning is, at its core, a measurement tool. And we must not make the grave error of confusing measurement with interpretation. In this case, given the history of biased policing practices in China (where the sample for the dataset was collected) that skew historical conviction data to overrepresent certain types of appearances,[10] there is a strong argument to be made against the compromised quality of these documents for establishing an adequate reference for a ground truth of facial criminality. Recognizing patterns is only meaningful if the data in which the patterns exist is also sound. In this case, the ground truth is not only *socially* problematic but also *technically* deficient, given the skewed data collection practices underlying it.

GTT is important precisely because of its ability to make explicit these inadequacies in ground truth construction. It allows for an understanding of which machine learning claims are more realistic (e.g., OCR) and which might be products of eloquent storytelling and weak methodological assumptions (e.g., criminality prediction from photos taken at the time of arrest). Although pernicious machine learning tools such as "criminality" prediction immediately raise red flags due to the high stakes associated with the name, other comparably subtler (but not necessarily unproblematic) applications such as predicting "employee fit" or "emotion recognition"—both of which are existing use cases I encountered in my fieldwork—may not. Performing GTT for such systems enables one to gauge its practical feasibility and understand its strengths and shortcomings without necessarily "opening up" its algorithmic "black box." After all, as many researchers working under the banner of "critical technology studies" have pointed out, algorithmic transparency is a limiting framework through which machine learning systems can be understood or interrogated (e.g., Burrell,

2016; Christin, 2020; Seaver, 2019). GTT provides an alternative tool that translates the problems of efficacy and opacity into a question of learnability, thus providing an explicit angle and site of interrogation: tracing the process of formulating the problem and scrutinizing the underlying dataset (including the materials through which it is comprised).

On the most practical level, this means foregrounding the *specific task* and *context* that the machine is calibrated for. Indeed, learning is, by definition, a practice that is always in relation to another process. Neither human nor machine never "just learns" but rather always "learns *x*." Understanding *x* is thus an indispensable component of interrogating machine learnability. In the context of voice AI, this takes the form of examining "how a machine is made to learn to listen to voices." Here, GTT more explicitly engages the ontological complexities of voice and the inherent variabilities of listening because it forces one to define voice analysis as a specific machine learnable task designed to do *x* (predict job performance, recognize emotions, etc.). It moves beyond conceptualizing the tensions between voice and particular technological forms (as we examined in chapter 1) and requires tracing a specific human process of problem formulation and dataset construction.

Case Study: Vocal Résumés

As part of my fieldwork, I encountered a startup that offers voice analytics for a suite of behavior prediction, mental health diagnosis, and personality profiling applications. One of the use cases for its voice analytics system was to predict whether an individual will successfully fit into an existing team for recruitment purposes using machine learning[11] (figure 2.2).

I spoke separately with the founder and chief scientist (Adam), a senior-level executive (Brian), and a mid-level manager (Colby) at this company.[12] I asked them questions around the design and efficacy of their system, some of which included what the voice signals are that would help predict employee fit, what the labeling process is like, and how voice data is collected. The responses I received ranged from a genuine "I don't know the specifics" (Brian) to "years of research have allowed us to correlate certain vocal features with certain behaviors and personalities" (Adam) and, of course, "there are limits to how much we can disclose the details of how our systems work" (Colby). Simply put, the responses I received were

What if you could hire candidates with the highest probability to succeed for a specific role, by simply analyzing their voice?

What if ... job and culture "fit" assessment was available to your Recruiters after just seconds of the first interview or screening phone call?

Focus on the Candidate

[redacted] enables talent sourcers to liberate your hiring teams from the tactical screening process. Our acoustic analytics empowers your recruiters to do their most valuable work by focusing on the candidates that matter the most- making hiring more intelligent.

Screening Made Easy

Our patented technology is calibrated to specific interaction patterns, personalities and behaviors that can be used to accurately screen your candidates early in the hiring process.

Gain Valuable Insights

[redacted] provides language independent, un-biased, objective pre-screening data that allows talent sourcers to focus their time and attention on those candidates most likely to succeed.

Figure 2.2
A screenshot from the voice analytics company's "Candidate Success Predictor" product page taken in January 2023. It details claims made by the company around the use of their AI voice analytics technology for recruiting purposes. The name of the company has been blacked out.

overall unproductive in helping me understand exactly how the so-called magic happens (Elish & boyd, 2017).

This is a common obstacle in conducting research on technologies primarily developed in proprietary settings. Indeed, machine learning practices in the tech industry are notoriously protected by nondisclosure agreements that make it difficult to conduct close tracings of their underlying infrastructures. As such, existing investigative methods that require direct engagement with ground truth datasets (e.g., Poirier, 2021; Van Rossem & Pelizza, 2022)—while undoubtedly robust when direct engagement is possible—are unfortunately not always applicable to studying these proprietary systems. GTT provides one way to circumvent these barriers.

As a method that foregrounds scrutinizing not only the underlying dataset but also the *process* of constructing it, GTT incorporates a retroactively deductive approach that starts from the end task to trace how a ground truth might be established for it. It can thus start from the underlying hypothesis—in this case, "employee fit is measurable from voice signals"—and work toward deducing the ground-truthing process. This is not to say that the most robust and ideal iteration of GTT would not require direct engagement with the underlying datasets; rather it emphasizes that even when faced with a relative lack of such information, the method can still inform a systematic "problem-oriented" (Jaton, 2021) perspective for deducing the qualitative processes through which ground truth datasets are created.[13] GTT can thus be thought of as an especially useful tool for situations in which access to datasets is limited and information is scarce—precisely the situation I found myself in with this voice analytics startup. Indeed, although it is likely impossible to know the exact details of how the particular voice-based employee-fit prediction model developed by this startup works without extensive cooperation from its employees, it is possible to deduce through GTT how the problem of learning the "vocality of employee fit" might be formulated: *How would employee fit be measured? And how would it be measured from voice?*

To address the first of this two-part question, so-called employee fit is presumably a state determined by the relationship between the potential employee, the members of the team for which the employee is being considered, as well as the work, objectives, and motivations that drive the team. The "team" would also have to be bounded as a specific group of people with varying personalities and skills that work together toward particular team objectives. Establishing an individual's so-called fit-ability as a measurable, comparable, and rank-able property is thus contingent on also being able to measure and quantify these other related and fluctuating sociopsychological variables. In this way, "fit-ability" resembles what is described in cognitive psychology as a *construct*: "an abstract, unobservable, hypothetical entity inferred from postulated thoughts and observable behaviours, representing patterns of psychologically related phenomena" (Gignac & Szodorai, 2024, p. 1).

Psychometrics is the subdiscipline of cognitive psychology concerned with the measurement of such psychological constructs. It is beyond the scope of this chapter to survey and examine the intricacies of psychometrics as a field of study, but it is worth noting that there are extensive operations through which psychometric constructs such as "intelligence" and "personality" are deemed "measurable." Psychometric intelligence and personality, for instance, are founded on more than a century of scholarship and research that also catalyzed the invention of some foundational statistical techniques such as factor analysis (Spearman, 1904). Even with this extensive history of research, however, there are still significant critiques around both the validity of these measures and their misuse as tools of social control (e.g., intelligence and IQ, personality and OCEAN).[14] In addition to these social critiques, another fundamental reason for this pushback is that the measures are entirely dependent on statistical correlations and proxy tests. In fact, a psychological construct is only considered a construct if the property to be measured is unobservable directly. It is *defined* by its immateriality, which is why correlations of other observable behaviors or testing outcomes must be recruited to "measure" the construct.

The company's approach to establishing "fit-ability" as a measurable property would rely on a similarly correlative/proxy logic. Indeed, not unlike how the abstract notions of intelligence and personality are deemed "measurable" through proxy tests that must be rigorously validated and tested for statistical reliability, establishing "fit-ability" as a measurable construct is also dependent on constructing quantifiable metrics for all of the aforementioned variables and more—for example, the team's unique cultural dynamics, interpersonal relationships between existing members, the interplay between potential members' various skills and those that are missing from or needed in the team, and so forth. As such, GTT shifts our attention from the abstract frame of "fit-ability" to the material processes of collecting and constructing these hypothetical metrics: What are the *materials* through which a ground truth for "fit-ability" is established? Are there self-reports of existing employees? Are there results of personality tests of existing employees? How has the team's unique culture been quantified? Are there skill profiles of existing employees? Are there measures that connect team profiles with team performance? In these ways, GTT allows more concrete questions to be asked that interrogate the course of developing a ground truth and transforming "employee fit" into a machine learnable problem. As Sadowski (2025) writes, "Ground truth is a way to transform hyper-specific segments of reality into a medium that is machine readable; thus making them legible to systems that are extremely limited in their capacity to interact with the world" (p. 7). GTT focuses on these "mediums" and "transformations" as crucial sites and processes for interrogating the machine learning systems built from them.

Addressing the second part of the original question regarding how employee fit would be measured from voice further emphasizes the near-impossible process of ground-truthing the *vocality* of employee fit. Even if one were to assume that all of the aforementioned proxies ("materials") for "fit-ability" were available, one would still have to accept that meaningful correlations between those proxies and specific vocal features extracted from existing employees recorded and measured across various situations exist. This is with the additional assumptive caveat that the presence of recording devices does not alter the ways in which the recorded individuals not only vocalize but also *feel*. In fact, considering the numerous situational (Eidsheim, 2018), physiological (Kreiman & Sidtis, 2011), and emotional (Scherer et al., 2003) variables that affect vocal expression, the attempt to correlate "fit-ability" measures to vocal features for the construction of a ground truth database would, on a practical level, border on impossibility. Even if it were assumed to be possible, it would still be an immensely tedious, expensive, and error-prone procedure, as well as an undesirable one with regard to generalized scalability, considering the uniqueness and variability of each team.

So, how is this company in business? The reality is that none of these variables are meticulously measured or even accounted for. Indeed, one of the primary values of a ground truth is that these separate variables are presumed to exist in the data itself: "The key properties of the ground truth—quality, veracity, accuracy, relevance, and so forth—are usually implied and rarely investigated" (Sadowski, 2025, p. 7). Here, the sequence of assumptions might go as follows:

1. It's probable that voices carry signals of personality.
2. Personality is presumably important for determining how well an employee will fit in with a team and perform in a job.
3. A collection of voice recordings of current employees should contain signals of their aggregate personalities and likely also other relevant signals in voice that haven't been specifically identified.
4. Voice recordings of existing employees can thus be a useful reference (ground truth) to compare the voice recordings of incoming potential candidates to.

It's an astonishingly simple—"brute force"—approach to "producing" knowledge. The underlying assumption is that enough examples should contain the relevant rules, so collecting enough examples can substitute the work of figuring out the rules.[15] As Campolo and Schwerzmann (2023) document, this "example-based paradigm" has become the dominant mode of computing in machine learning. Indeed, a significant motivation for constructing ground truths is that they can enable action without explanation. Under this frame, anything and everything for which data can be collected can be deemed machine learnable. The objective is not to critically address the intricacies of the problem (as we did above) but rather to *assume* the relevant relations exist in the collected data.

Of course, this is not always problematic. As philosopher Michael Polanyi (1958) writes, a significant part of our practical knowledge is indeed unproblematically *tacit*: Most walking individuals, for instance, are not aware of the specific rules that govern their ability to walk, and this does not present a practical problem for their walking. Even in humans, a certain degree of tacit knowledge is always expected in any form of expertise and action. In the context of machine learning, too, there are reasonable situations in which the assumption of relevant relations alone is adequate and understandable as a means of establishing a ground truth. For instance, a computer vision model designed to recognize images of dogs that is trained on a ground truth dataset that contains billions of images of dogs in various settings and styles is a reasonable configuration. The model is designed to recognize images of dogs, and the ground truth dataset contains many images of dogs. The ontological dissonance between what the model is designed for and what is included in the ground truth dataset is minimal. This minimal dissonance means we do not necessarily need to parse the specific features through which the model is able to complete the recognition task.

In contrast, a model designed to determine how well an employee will fit into a team based on voice signals requires significantly increased levels of mediation and interpretation. A massive leap is required to interpret an "audio file of speech" as an adequate reference for "how well a person will fit and perform in a team." Here, theoretical explanations become necessary to justify this interpretation. Structurally, it is a wholly different schema compared with training a model to recognize the exact "thing" that it was trained for (images of dogs). Here, the system is dependent on an assumption not only of *features* (i.e., that there are certain characteristics that represent a dog, which is reasonable) but also of hypothetical *interpretations* (i.e., that voice recordings contain accurate and reliable signals for how well a person will fit into a specific team, which has never been scientifically validated). Indeed, a surprisingly underutilized frame for examining the feasibility of machine learning models is the intended output. If the intended output is what is contained in the ground truth (images of dogs), one can reliably presume that the model will reproduce that output (recognizing dog images). In contrast, if the intended output ("employee fit") is not directly observable in the ground truth (a collection of voice recordings), then one should be skeptical of the degree to which the intended output will indeed be output*ted*.

Studies of algorithmic technologies that analyze résumés, for instance, have provided important insights on such limitations. As Schellmann (2024) documents, the ground truth of résumé screeners for incoming applicants are similarly composed of résumés from already hired (interpreted as "successful") applicants. In one investigation, "having played high school lacrosse or being named Jared" (p. 30) was considered a primary predictor of performance. In another, having "baseball" on a résumé was evaluated positively, whereas having "softball" was scored negatively (p. 30). In both cases, the likely explanation was simply that these were correlations found in previous applications of hired individuals, which comprise the ground truth. In other words, a model developed from a ground truth that is constructed from previous résumés is simply a model that "predicts" how similar incoming résumés are with previous ones. "Success" and "fit-ability" are *interpretive* variables that human designers project onto the ground truth, but the ground truth itself is not directly representative of these interpretations. Of course, this is not to say that there is a *zero* percent chance that these correlations are meaningful, but the burden of proving that these correlations are not mere statistical artifacts and indeed meaningful signals of high performance must be with those who deploy these tools and not those who are subject to their assessments. Unfortunately, this is rarely (likely never) the case.

The same can be said for the voice-based employee-fit prediction system I encountered. According to Adam, the technology is designed "not for a specific use case. It is for a more generic platform that collects the vocal input and then uses predictive analytics to correlate vocal features and patterns with different behaviors and personalities."[16] Adam also asserted that they are developing technologies that are not only "language independent" but also "culturally independent," reiterating the widely observed but also heavily critiqued folk theory that the timbral qualities of voice/speech (as opposed to semantics) are stable and direct representations of an individual's "authentic" inner state.[17] In Adam's own words, "Voice is a direct link to your character; it can't be manipulated." Compiling Adam's claims, this system is not tied to a specific use case, language, or culture—it is dependent on the assumption that there are universal vocal features that have been identified in relation to "fit-ability" and captured by a model that can simply be fine-tuned for particular use cases ranging "health, financial behavior, consumption patterns, and interpersonal communication tendencies" (Adam). In reality, it is simply a model that measures similarities of vocal features between the ground truth and the new input.

Despite the obvious *technical* shortcomings and unvalidated assumptions underlying such an approach, it does provide *practical* benefits for a company that uses the tool. From a financial standpoint, the cost-saving advantages of automated approaches to recruiting and applicant screening are often deemed by administrators to outweigh the missed opportunities of identifying the best applicants (Schellmann, 2024). Indeed, within a capitalist frame, all business activities can be explained through profit, which means recruiting is not necessarily about hiring the best candidate but always in service of the underlying goal, which is to contribute to the company's bottom line. A recruiting tool that saves costs but misses "fitting" employees can thus be seen as a desirable alternative to traditional and vastly more expensive hiring practices if the financial result is net positive.[18] Indeed, the recruitment tool "works well" if "working" is defined through profit.

What is important here is that gauging learnability via GTT makes the materiality, and thus also the technical boundaries, of "what exactly is learnable" explicit. If a ground truth defines the range of a machine learning model's possible outputs, then tracing the process of ground-truthing a problem—"predicting employee fit from vocal input"—sheds light on the process of materially translating the problem into data. As I show, however, this translation process is dependent on a series of assumptions, ontological reductions, and importations of external standards that not only change the form of the machine learning task but also introduce new problems, both social and technical. Indeed, like the facial criminality prediction model by Wu and Zhang (2016a), the "ingredients" of the ground truth provide concrete sites for critique. In the same way that machine learning practitioners substitute theoretical explanations with data, critical scholars or those who seek to challenge the claims of this system don't need necessarily to "prove" that "fit-ability" is immeasurable from vocal signals. They simply need to show that the ground truth for vocal "fit-ability" is insufficient: There is too much ontological dissonance between what is materially contained in the ground truth and the interpretive claims that are made using models developed from it.

I thus present this example to bring attention to the granularity that a potential use case requires a machine learning system to account for. It is precisely because of this specificity that current machine learning systems cannot be discussed in separation from the contexts in which they are deployed—and also why claims around machine learning systems that seek to extend well beyond a particular context should be met with skepticism.

Subjectivity, Stakes, and Power

The GTT/learnability framework is a useful analytic tool to gauge the practical feasibility of a machine learning system, but what is missing or unclear in its application are discussions of subjectivity, stakes, and power. Although it explicitly foregrounds the question of how this is made possible, it does not always engage questions about who this serves, what the stakes are and for whom, and whether it *should* be made possible.

In attending to these questions, I momentarily return to Neisser's 1963 essay, "The Imitation of Man by Machine." His differentiation between explicitly "technical applications" and "social decision-making" is an important one for understanding not only the epistemological limits of machine learning systems but also the ethical ones. Contextualized in a GTT/learnability framework, the difference between "technical" and "social" can break down in the most basic sense to:

(a) Technical: a problem or task in which the spectrum of possibilities is bounded, and performance or "success" is defined explicitly.

(b) Social: a problem in which the spectrum of possibilities is unclear, constantly changing, and contingent on broader social factors that result in a fluctuating, interpretive, or contestable standard of performance or success.

Sorting the various examples mentioned in this paper according to these two criteria makes evident the utility of the framework in gauging the applicability of machine learning to a particular task. OCR, for instance, fits relatively neatly into category (a), whereas tasks such as saliency detection or employee fit fall under category (b) but can potentially be reconceptualized—generating varying degrees of ontological dissonance—into a problem that fits category (a). Only once a problem is "acceptably" fitted to category (a) can it be "solved" with ML. In other words, one key practice of machine learning practitioners is to conceptualize problems, so they fit category (a) and not category (b).

Rachael Tatman,[19] an expert in NLP with a PhD in computational linguistics whom I interviewed in the summer of 2022, refers to category (a) as the "boring" problem, in her words, "like plumbing." She presented this simile not to view plumbing as a vocation or an infrastructure that is by any means trivial or uninteresting but rather to express her hope that machine learning can also one day be understood as an important but somewhat taken-for-granted substructure that facilitates specific straightforward tasks. This is a deliberate discursive positioning against the sheen and hype that uplifts and drives so much of the current artificial intelligence industry. Expressing her concerns around the absence of a "standard of care"[20] for ML engineers, Tatman was particularly skeptical and worried about machine learning systems that "make decisions about people," which almost always fit in category (b). As I emphasize throughout this chapter, such machine learning tasks (i.e., Wu and Zhang's criminality-prediction system) are exposed of their *technical* shortcomings when examined through a problem-oriented perspective (Jaton, 2021) afforded by GTT. Ethical concerns, however, are often left unaddressed.

This is vital, especially in the lack of regulatory guardrails. There are few limits that require ML engineers or the broader organizations for which they work not only to acknowledge the deficiencies of a particular system but also to foreground a system's capacity for harm. Tatman pointed out that this is exacerbated by the fact that the people whom the decisions are being made about by these machine learning systems "don't have legal or practical power over those decisions." Understood in relation to automation bias, which states that people are more likely to believe automated decision-making systems and to ignore contradictory information made without automation, even if the nonautomated information is correct (Skitka et al., 1999), using machine learning for making decisions about people ultimately means that power is consolidated among those deploying and using the systems, whereas it is taken away from those it is being used on.

Such widespread lack of understanding around machine learning applicability and techniques exacerbates a corporate environment in which "spinning a good enough story will secure funding" (Rachael Tatman). A tech culture in which overly ambitious projects and magical thinking can command immense capital and human resources (e.g., Elizabeth Holmes and Theranos) contributes toward the increasing attraction toward provocative narratives while moving us away from engaging with the complex realities and true potentials of machine learning.[21] As Tatman emphasized, there is an urgent need to engage both the practical possibilities of machine learning and the broader stakes of its applications.

David, the principal scientist at a "unicorn"[22] conversation analytics company at the time of our interview, shared similar concerns. He was adamant in his view that machine learning research and implementation should always "stay away from high-stakes industries such as health care or law enforcement." According to David, this was one of the reasons he chose to apply his PhD training in machine learning to what he deemed a "lower-stakes" context such as customer service. David's story, which elaborates upon many of the concerns introduced by Tatman, presents an interesting example from which we can think through the interplay between a GTT/learnability framework and a subjectivity/stakes/power one. Or more specifically, how the aperture of subjectivity/stakes/power can change how one might understand a machine learning model that GTT might show to be ontologically consonant with the qualitative phenomenon it is modeling.

Broadly speaking, David's work consists of using machine learning techniques to identify signals that are indicative of "sentiment," "intent," and "flow" of a conversation. Constructing a ground truth for a system that identifies "intent" is of course extremely difficult. It consists of initially collecting massive amounts of data of previous calls with known outcomes as well as drawing heavily from existing NLP literature and vocal expressions of emotion/sentiment to make correlations between specific linguistic patterns and audio signals with variously identified relevant "intents" (i.e., "refund", "exchange", "balance check"). GTT makes explicit the challenging and contextual nature of this process, but because, in the case of the customer service industry, large internal archives of recorded conversations do exist, it is not necessarily impossible to find somewhat usable correlations between conversation patterns and predefined and straightforward actions that follow (interpreted as "intents").

That being said, David shared that the accuracy of predictions is still at a very low level—in his words, "slightly better than chance." If the goal is simply to add a layer of analysis to help streamline incoming customer service calls, he said the technology can help. With regard to finding reliable and accurate signals that will consistently be able to recognize ambiguously defined concepts such as "intent," however, the research is nascent. Following this, he contrasted his work with that of software engineers and emphasized that there is minimal structure to how a machine learning research team works because it's not about "fixing a specific problem or adding a feature." It starts from a more fundamental state of turning something into a problem and then experimenting with machine learning to see if that problematization is the most efficient or even the correct approach. He mentioned that because of this, a machine learning research team never truly knows if it's on the right path. Mentioning the lengthy research periods and copious amounts of resources required for research and development in machine learning, David shared that the time and capital could arguably be better used for other purposes.

These resources, of course, refer not only to the vast amount of computing power required to train machine learning models but also to the global infrastructure of underpaid human annotators required to construct, maintain, and make usable the ground truth training data that scientists such as David work with (Elliott, 2021; Hao & Paola Hernández, 2022; Wang et al., 2022; Williams et al., 2022). In this way, examining machine learning systems through the aperture of ground truths also explicitly foregrounds the colonialist underpinnings (Hao, 2022) of the AI industry by shifting attention away from the sheen of algorithms, engineers, and industrial research labs and toward the actual annotators (often in the Global South) who do the laborious work of building the ground truth datasets upon which all machine learning systems depend.

David also voiced his concerns around the lack of conversations AI as a Service companies in general have with the people they say they serve. In the context of his conversation analytics company, he expressed that augmentation and surveillance are two sides of the same coin. Because the machine learning team requires mining data from the conversations of customer service calls, there is constant audio monitoring of customers and call center agents. Not only is this a concern with regard to storing sensitive information and surveilling agents' actions, but it also means that in efforts to formalize a task (customer service), the machine listening system is exploiting and commoditizing data gathered from the craft of customer service agents. Mentioning that there are very few standardized evaluations of the concrete benefits of these technologies, David said he is concerned that he is contributing toward a system that is sold based on a story of helping workers but which ultimately has the potential to make fungible those same individuals. After a pause, he said that he likes the general idea of empowering people with tools as long as there is sufficient dialogue and understanding between developers and users.

David's perspective is important because it both highlights and complicates the utility of the GTT/learnability framework. In his discussion of choosing customer service over other industries such as health care or law enforcement, he expressed the significance of considering what the *stakes* are for the machine learning work that is being conducted, regardless of how *feasible* it is. This continues in his contemplation on whether the machine learning tools he builds will actually help the individuals they were intended for and, further, if the resources required to sustain ML research and implementation are justifiable. It thus broadens the perspective from an ontological and material interrogation of groundtruthing practices to a social and ethical consideration that attends to the varying power dynamics that mediate the network of actors upon which machine learning is dependent. Filtering the two aforementioned dimensions of GTT/learnability through the degrees of "stakes" ultimately narrows down the range of applications for which machine learning is *ideal*.

Acknowledging that both GTT/learnability and "stakes" are better expressed as spectra as opposed to discrete cells, table 2.1 presents an intentionally simplified visualization representing how these questions around subjectivity, stakes, and power mesh with and change the GTT/learnability framework. As is evident, the most ideal context is currently one in which there is a relatively bounded standard, which means that high-quality ground truth labels can be developed that do not require extensive translations from the original problem, *and* the stakes of both the conditions for ground truth construction as well as the effects of the machine learning system are low. Not only that, but this orthogonal analysis also complicates previously more "clear-cut" situations: A situation with a fluctuating standard might not always reflect a deficient machine learning system if the associated stakes of development and output are low, whereas a situation with a bounded standard might not always reflect an acceptable machine learning system if the associated stakes of development and output are high.

Table 2.1
Orthogonal relations between stakes and GTT/learnability

	GTT/Learnability	
Stakes	(a) Bounded Standard	(b) Fluctuating Standard
Low Stakes	√	X→?
High Stakes	√→?	X

In the same way that Jaton (2021) discusses the costs and benefits of ground-truthing practices, however, this simplified table above should not be taken as an end-all index against which the possibility and utility of all machine learning systems are measured. There are, after all, many highly debated cases in ML research such as autonomous driving, for which the stakes are extremely high, with its standards of learnability ranging anywhere from extreme variance to partial boundedness (depending on the metric by which one considers an autonomous driving system "successful"), that many believe drive rigorous academic and industry research. This table is meant to serve as not a shortcut for validating or discounting such work but rather the opposite. It is to provoke further contemplation and to introduce additional dimensions through which one may explicitly foreground the necessary ontological translation processes that are required to make a problem practically learnable, as well as consider the broader social questions around who/what is at stake and who/what that serves, with the ultimate purpose of more effectively engaging with and understanding a machine learning system.

It should also not be mistaken as any kind of groundbreaking addition to the field of machine learning as it represents a perhaps previously tacit but still prevalent thought process by which any ML engineer or scientist must approach a machine learning problem. Instead, taken together with the perspectives presented in this chapter, it should be understood as a strategically reductive heuristic through which ML engineers, data scientists, social scientists, humanists, activists, policymakers, and even business analysts interested in artificial intelligence can collaborate, debate, and think about such systems together.

Ultimately, this learnability–stakes (LS) table is my response to a tendency I observe among some humanists, social scientists, and, perhaps more importantly, policymakers/analysts to express social concerns around technologies with minimal discussion or understanding of their technical capacities, which I believe is ultimately not very different from the overly celebratory hype culture of the tech world (only from the other side of "fear"). The GTT/learnability framework is meant to act as a funneling tool to help non-technicians better understand the operational logics of machine learning and make explicit the cascade of decisions that massage a problem into a machine "learnable" one. On the other side, this LS table is meant to serve as a push for the developers working in the broader machine learning community to more explicitly foreground questions of subjectivity, stakes, and power that humanists and social scientists have brought forth in engaging with these systems. As I mention throughout, the eventual hope is to build a robust foundation for effective interdisciplinary dialogue and collaboration.

* * *

I mentioned at the end of interlude 1 that this chapter would address how machine truths are constructed—a detour and foray into the world of machines and how they are built. Despite this precautionary notice, the relatively sparse mentions of "voice" and "listening" in this chapter may still remain a question mark for some readers. This curiosity is not lost on me. And my short answer is that the absence is ultimately a reflection of the data.

I found throughout my fieldwork that although everyone I spoke to worked or was working in the voice AI industry, the complexities of voice and listening were sidelined by discussions around the business potentials of AI/ML for leveraging "voice" as a previously underutilized domain of data analytics.[23] At best, the complexities of voice and listening were selectively grappled with as technical wrinkles that needed to be smoothed out into a machine learning problem. For the majority, however, folk theories of voice and listening (as opposed to more robust scientific and/or philosophical engagements with them) or what Eidsheim and Meizel (2019) refer to as "voice and its imaginary" appeared to be sufficient in informing the industry-wide justification around the importance of voice as the "next frontier" of AI and machine learning.

Under the framework of machine learning, all qualitative phenomena break down into patterns and correlations. In the words of Mackenzie

(2017), vectorizing data "produces a common space that juxtaposes and mixes complex localized realities . . . Similarity and belonging no longer rely on resemblance or a common genesis but on measures of proximity or distance, on flat loci that run as vectors through the space" (pp. 73–74). Quantitatively flattened into vector space, "voice," "listening," "intent," "criminality," "employee fit," "saliency," and so forth are all translated into correlatable data. Of course, not all translation processes are equally smooth, and examining each type of application through the lens of GTT/learnability makes explicit the degree of ontological dissonance that is produced in such translations. Based on this, I sought to show that the most useful applications for not only machine listening technologies but also machine learning systems writ large are those in which the range of possible outcomes are bounded (i.e., games such as chess or *Go*, recognizing characters, detecting objects, etc.) and the implied stakes of both the conditions for development and the various outcomes are low. This is not to say that other uses that don't entirely conform to these qualities, such as most generative AI models,[24] are deficient. But rather, it is to present a frame of thinking that allows one to qualitatively engage with machine learning systems in systematized and grounded ways. In this sense, GTT and LS should be seen not as heuristics of simplification but rather as heuristics for *complicating simply*. Indeed, we often hear the phrase "there is no right answer" to many of life's most profound questions, and there is a beauty in that uncertainty. We should not try to find answers to those questions with machine learning. Rather, machine learning might be best suited for those "boring" questions and bounded tasks, where there may in fact be a "right answer."

In the following chapter, I shift my attention to a sub-domain of machine listening—speech emotion recognition—in which there is an explicit *lack* of scientific consensus on an agreed upon ontology of emotion. There is *officially* no "right answer," and this lack is precisely what makes this domain both problematic and interesting. Indeed, I will show how GTT as a method can help us trace the processes through which emotional voicescapes are constructed for the purposes of using machine learning to detect emotions from speech. If this chapter was a detour through the world of the machine, chapter 3 brings us back to the world of voice and emotion.

3 Listening to Engineered Emotions

The Sound of Emotion

"Let me tell you something."

41 percent fear + 14 percent anger + 14 percent sadness + 7 percent confusion + 7 percent distress + 3 percent amusement + 3 percent disappointment + 3 percent ecstasy + 3 percent embarrassment + 3 percent surprise (positive)

In 2019, the science journal *Nature Human Behavior* published a study that examines humans' abilities to detect emotion from speech (Cowen et al., 2019). According to the study, of the eleven emotions that apparently comprise the utterance "Let me tell you something" by speaker 148 of the Vocal Expressions of Nineteen Emotions across Cultures (VENEC) corpus (Laukka et al., 2010),[1] fear is the most indicative of this speaker's emotional state. Indeed, when one listens to the vocal sample, a likely assumption might be that the speaker does not sound particularly happy. Whether the tone is more indicative of frustration, irritation, confusion, or fear, however, is contestable. Presumably, one's perception or classification of a vocal expression as sonically representing a particular emotion would depend on personal experiences that allowed one to form mental associations between certain tonal features of speech with particular emotional signals represented by learned emotion labels. The sound of fear for one person can vary significantly from that of another.

So, what *is* emotion, and how does one know or, in this case, *hear* it? To continue the example of fear, if it is an emotional construct that varies from person to person, it also means that its boundaries are not definitive but rather fluid. Where does fear begin, and when does it stop? Which

41 percent of the utterance by speaker 148 constitutes fear? Does the fact that 41 percent of the annotators perceived the utterance to reflect fear ascertain the sample's audible fearfulness? Which part of the utterance correlates with the 3 percent of ecstasy? Or the 14 percent of anger? Or the 3 percent of amusement? The absurdity of these questions is natural, and reflective of the ambivalence that characterizes what exactly one means when referring to "emotion." Indeed, as psychologist James A. Russell (2003) writes, "There are no formal criteria for what is and what is not an emotion" (p. 145). If anything, he suggests that the term "emotion" should represent a general banner under which different events, states, reflexes, sensations, feelings, and attitudes, among others, can be generally discussed, without necessarily having the power to denote ontological boundaries—both in the sense of categorizing particular phenomena *as* emotions as well as in the sense of differentiating between what *is* and *isn't* an emotion.

This deconstructive move by Russell is primarily a response to the theory of "discrete emotions,"[2] which uses labels such as "fear," "anger," and "joy," among others, to refer to basic and biologically determined states that are posited to exist across all humans. Influenced by Charles Darwin's (1872) writings in *The Expression of the Emotions in Man and Animals*, psychologists such as Silvan Tomkins (1962, 1963) and, later, Paul Ekman (1970) popularized the notion that prototypical and universal emotions can be traced back to evolutionarily developed behaviors and physiological reflexes.[3] Despite the popularity of the theory, however, numerous scholars have pointed out that this attempt to reliably link emotional experiences to physiological patterns was never fully realized (e.g., Dror, 2001; Russell, 2003). Although there is indeed evidence that points to correlations between emotion, physiology, and behavior, no scientific convergence exists on the exact contours, strength, or reliability of those correlations.

To return once more to the earlier example of fear, consider the well-known "bear" example provided by psychologist William James (1884). Here, James theorizes that the evolutionary act of physically fleeing from a predator in the wild leads to the psychological sensation of fear (emotion). It is not the bear but rather the psychological sensation of fleeing from danger that elicits the emotion. Of course, as Russell (2003) points out, this fear is not the same emotion that one might experience watching a horror film. There is no danger, there is no escape, and the experience is arguably enjoyable for the person paying to watch the film: "What, other than the label *fear*, do various instances of fear share with each other that they do not share with what is not fear? . . . There may be no one scientific model that applies to all cases of fear, and only to fear" (p. 146).

These complexities of vocalized emotion, or, more specifically, vocalized emotion *science*,[4] are crucial mediators in conceptualizing, designing, building, and deploying any technological iteration that might be housed under the banner of "speech emotion recognition."[5] As I elaborate in chapter 2, any machine learning model that is made to recognize, detect, and classify qualitative phenomena can only do so through an internal translation (ground-truthing) of that qualitative phenomenon into a statistically legible proxy. In an SER system, dynamic and complex concepts such as speech, voice, and emotion must all be stabilized through such truth labels. This means that these ontologically multiple phenomena need to be not only bounded but also rendered epistemologically coherent: Emotion is *x*, and it can be known (or heard) through *y*.

In this chapter, I apply the methodological framework introduced in chapter 2, ground truth tracings, to show that these methods of knowing, classifying, and stabilizing in machine learning SER applications invoke particular ontologies of emotion and voice while ignoring or inadequately engaging with others, and, in the process, require *redesigning* and *reproducing* a reduced "emotion-fit-for-ML" proxy as SER ground truth training data.[6] This chapter thus demonstrates how GTT can leverage the artifacts created for/around a machine learning system such as datasets and methodology papers to empirically trace the construction of ground truth proxies. In this way, I take what Jaton (2017, 2021) refers to as a "problem-oriented approach"[7] to examine the steps through which "emotion" is simplified, stabilized, and bounded—by engaging directly with the training datasets that taxonomize emotion and the accompanying papers that detail the processes through which emotion is "fastened" to the "format" (Koopman, 2019) of a training dataset.

As philosopher Colin Koopman (2019) writes of "traits" in personality, "Once a unit is rigorously standardized and portable tools for its measurement are furnished for use, the unit measured by the standard *becomes real*" (p. 85). Just as the kilogram was made real through the development and maintenance of the metric system and the International Prototype of the Kilogram (Mulvin, 2021), emotional constructs such as the "voice of anger" and the "voice of disgust" *become real* through the technologies that purport to detect them. As these technologies are leveraged to make important decisions such as determining how a worker is evaluated and even compensated—which, as I will show in the second half of this chapter, is indeed how call center agents are treated through the frame of "call center optimization"—there emerges an urgent need to scrutinize the methods through which that realism, or what Campolo and Schwerzmann (2023) refers to as "artificial naturalism," is retroactively constructed.

Fitting Emotion to Machine Learning

"We are constantly developing more accurate methods of measuring human emotions . . . It is possible, then, to speak of the emotion quantitatively, as being present in large or small amounts" (as quoted in Dror, 2001, from "Students Measure Fear," 1925).

The concept of emotion as a quantifiable and measurable object is an appealing and provocative one from the perspective of science. As historian and medical anthropologist Otneil Dror (2001) recounts, the numericization of emotion, or what he calls the new "*emotion-as-number*" (p. 359), was a powerful construct and stimulus in the "scientification"[8] of emotion during the late nineteenth century. "Specific physiological patterns signified an emotion . . . [it] was a pattern written in the language of the biological elements that one monitored in, or sampled from, the organism—translated into a sequence of numbers" (p. 362). In 1978, for instance, Ekman, one of the central figures in the development of discrete emotion theory, and his collaborator Wallace Friesen operationalized a schema called the Emotion Facial Action Coding System (EMFACS). This schema denoted various "action units" (AU) that described different muscles or muscle groups that contributed to facial movements. In Munn's (2020) words, the underlying assumption that upheld this "calculus of emotion" was that "emotion equals muscle one plus muscle two" (p. 25).

This scientific bridge constructed between emotion and physiology, of course, was indicative of a broader "body-as-machine" schema that permeated conceptions of human behavior and body at the time.[9] Similar changes were also observable in changing ontologies and epistemologies of voice during the late 1800s, when craniometry was thought to be a newfound "solution" for justifying racial differences in singing capacity and vocal timbre (Eidsheim, 2015). Similar to how a calculus of emotions depended on a modular-physiological view of emotion as an aggregation of different muscles, a calculus of voice drew from such bodily measurements such as skull size and shape. In this way, the inclusion of concepts such as emotion and voice into the common scientific domain of physiology was believed by its proponents to produce new correlatable features through which they could be described.

There were many attempts during the late nineteenth and early twentieth centuries that drew from such physiological understandings as the bases for the development of "emotion-reading" machines. Some of these included the "Emotograph" of the 1920s, which resembled "a small radio with dials and tubes and a sort of stock market tape at one end to record the emotions" ("The Hidden Truth," n.d.). There was also the "Emotion-Meter" developed by the head of Paramount Studio's sound department, Loren L. Ryder, "to record the spectator's heartbeat and rate of breathing as scenes of love, violence and excitement unfold[ed] upon the screen" ("Emotion-Meter for Use at Film Previews," 1946). Although these inventions were based on tenuous scientific links assumed between physiological phenomena and emotion, they were instrumental in and reflective of the broader cultural transcription of emotion into the scientific language of numbers and graphs. As Dror (2001) writes, "The number was an important technology for the reframing of 'emotion' and its integration into the discourse of the laboratory" (p. 371).

It was not until the turn of this century, however, that pattern recognition techniques—that is, machine learning—were applied to "classify *utterances* according to their emotional content" (Dellaert et al., 1996, p. 1970, emphasis added). With this, the premise of an emotion-*hearing* machine, as it is understood today in the form of SER, started to gain traction in the scientific community. Outlined in their paper "Recognizing Emotion in Speech," Dellaert et al. "recorded a corpus containing emotional speech taken from the believable agent domain [1], of over 1000 utterances from several different speakers" (p. 1970). It is important to understand here what the authors mean by "believable agent." This is a term coined by computer scientist Joseph Bates (1994) in which he draws from the work of Disney animators Frank Thomas and Ollie Johnston (1981) and their philosophies for expressing emotions through animated characters. Bates adopted these animation tactics as the bases for constructing self-animating creatures or "believable agents" called "Woggles." The three tenets that Bates took from the Disney animators were:

1. Emotional states must be unambiguously defined so that viewers can attribute definite emotional status to the character.
2. Emotions must be directly apparent in the action of the character.
3. Emotional expression must be accentuated through foreshadowing and exaggeration, even if it requires toning down other simultaneous actions because viewers cannot grasp emotional states immediately.

The speakers recruited for Dellaert et al.'s (1996) study were given similar guidelines in the way they were asked to recite sentences accompanied by one of four emotion labels among "happiness," "sadness," "anger," and "fear." In other words, a particular *theory* of emotion, namely one that subscribes to the notion of "discrete" emotions, as well as a particular *method* of emotional expression, one that is aimed at maximizing at all costs the perceived identifiability of a discrete emotion, were inscribed into the conceptual fabric of SER research from its very inception.

Indeed, when Rosalind Picard (1995) first mentions the concept of computers recognizing emotional speech in her field-defining text, *Affective Computing* (published a year before Dellaert et al.'s first actual application of pattern recognition techniques to SER), she draws from similar frames. Specifically, Picard includes a table (table 3.1) from the Acoustical Society of America (Murray & Arnott, 1993) that summarizes "the vocal effects most commonly associated with *five basic emotions*" (p. 179, emphasis added).

Despite the relative simplicity of the table, it should be noted that Murray and Arnott take great care in their original paper to engage with the different theories of emotion, including those of non-categorical models. Their table, however, is a product of an expansive literature review of existing studies around vocal features and emotion at the time, and because the majority of these studies organize emotions across the five categories labeled in the table, there is an absence of non-categorical representations of emotion.

Picard's (1995) adaptation of this table as a response to the question she self-poses—"What is a good computational mapping between emotions and speech patterns?" (p. 179)—is reflective of a common practice in machine learning, in which data, frameworks, and theories are imported from relevant "external" domains as ground truths. The issue

Table 3.1
Summary of human vocal effects commonly associated with the labeled emotions as found by Murray and Arnott (1993)

	Anger	Happiness	Sadness	Fear	Disgust
Speech rate	Slightly faster	Faster or slower	Slightly slower	Much faster	Very much slower
Pitch average	Very much higher	Much higher	Slightly lower	Very much higher	Very much lower
Pitch range	Much wider	Much wider	Slightly narrower	Much wider	Slightly wider
Intensity	Higher	Higher	Lower	Normal	Lower
Voice quality	Breathy, chest tone	Breathy, blaring	Resonant	Irregular voicing	Grumbled, chest tone
Pitch changes	Abrupt, on stressed syllables	Smooth, upward inflections	Downward inflections	Normal	Wide, downward terminal inflections
Articulation	Tense	Normal	Slurring	Precise	Normal

Descriptions are in relation to what the authors call "neutral" speech. Table is adapted from Murray and Arnott (1993).

with such importations, however, is that once a ground truth for a particular machine learning task is established and accepted, a research community forms around it through which that ground truth itself becomes the basis for the development and testing of further evaluation benchmarks and performance metrics.[10] This further entrenches it as the norm for how that task is conceptualized and experimented with (Jaton, 2021). Indeed, these assumptions, coincidentally informed by a philosophy of animation (a historical detail evidently buried in the scientific literature, given the absence of mentions in subsequent SER research papers) and a turbulent research community around theories of emotion in the field of psychology, are taken as the reference from which a ground truth is constructed in the vast majority of highly cited SER research papers published in the decade or so that follows (e.g., Hu et al., 2007; Nogueiras et al., 2001; Nwe et al., 2003; Schuller et al., 2003, 2004).

Although the influence of both Dellaert et al.'s (1996) study and Picard's (1995) book in adopting a discrete theory of emotions and a modular view of speech/voice as ground truth cannot be ignored, it would be inaccurate to say that their initial adoption alone explains why competing ground truths were not eventually established. Indeed, both a categorical view of emotion as well as vocal features described through acoustic dimensions such as pitch and frequency present useful taxonomies and measures within which the statistical processes of machine learning can be grounded. A categorical view of emotion, which posits that all emotional experiences can ultimately break down into around six discrete states (the number varies based on the particular theory, but the premise of separate "buckets" remains unchanged), presents a convenient framework of emotional taxonomy compatible with the classification logics of machine learning.

Indeed, as we discussed in chapter 2, this taxonomic representation and the identification of organizing labels are *required* processes in ground-truthing a qualitative phenomenon as a machine learning problem. There is thus a practical advantage in machine learning to maintaining a categorical view of emotion as opposed to adopting a dimensional (Russell, 2003) or constructivist (Barrett, 2017) model. Both emphasize (to varying degrees) the fluidity of emotion as a phenomenon that is contingent on changing situations and individuals, and thus difficult to stabilize/organize as "biologically hardwired" or ontologically stable. The alignment between discrete emotion theory and machine learning applications of emotion recognition can thus be seen as a utilitarian one, strategically prioritizing formal compatibility over scientific depth.[11] Emotion is "fitted" to machine learning, not the other way around.

Discrete Emotion Theory Versus Core Affect

A 2019 study by Cowen et al. (the study referenced at the beginning of this chapter), however, advances a stronger claim with regard to the *scientific validity* of discrete emotion theory in SER. Here, the authors set out to analyze "what drives the recognition of emotion, emotion categories (for example, Awe or Fear) or broader scales that capture core affect appraisals (Valence and Arousal)" (p. 369).[12] It is important to scrutinize their approach to this question because it specifically addresses a gray area in previous SER research studies, which have largely glossed over the lack of scientific consensus on emotional ontology and have instead relied on community norms around categorical views of emotion as ground truth for evaluation and performance benchmarks. Indeed, machine learning domains, such as "emotion recognition" and the communities that form around them are generally framed around "challenges" or "problems" to be solved (Orr & Kang, 2024). This means there is a key problematization process of turning a messy qualitative phenomenon into a usable quantifiable proxy. Once these communities are formed around such problems, however, the specific problems are largely unchallenged and taken for granted as representative of the broader "problem space" that they represent. By explicitly framing their study around a scientific comparison of what actually "drives the recognition of emotion" in the context of SER, Cowen et al. undertake an ambitious project to move beyond a *utilitarian* framing of the "emotion-as-category" problem to a *scientific* justification for it.

To test this, Cowen et al. (2019) recruited 2,345 participants from the United States and India and asked them to "judge at least 30 randomly selected speech samples from the VENEC corpus of 2,519 speech samples" (p. 370). The participants were randomly assigned to one of two response formats: One group was asked to select one of thirty emotional labels such as anger, embarrassment, fear, and surprise to evaluate the emotion expressed in the speech sample; the other group was asked to evaluate speech samples based on twenty-three different affective features (see Cowen et al., 2019, Supplementary Table 2) using nine-point Likert scales. Some of the questions that the participants were asked in relation to these affective features included those inspired by componential models (Scherer, 2009) such as "To what extent does the speaker feel like his/her situation is compatible with his/her self-image?" or "To what extent does the speaker feel like he/she can adjust to his/her circumstances?" in addition to questions pertaining to more traditional dimensional models (Russell, 2003) such as "To what extent does the speaker feel pleasant?" or "To what extent does the speaker feel stimulated?"

The research question and study design can essentially be seen as a response to Russell's (2003) theory of "core affect." Here, he dismantles the categorical view of emotion based on its lacking capacity to establish a consistent method and scientific ground for defining what is and isn't an emotion. Highlighting that many scientific studies of emotion actually examine different phenomena spanning "brain modes, actions or action tendencies, reflexes, instincts, attitudes, cognitive structures, motives, sensations, [and] feelings" (p. 145), Russell proposes that for the *scientific* study of emotion—an amorphous and composite construct—there needs to be a universal and irreducible unit to describe emotion*al* experiences. He thus proposes the core affect model, a dimensional framework that orthogonally maps valence (positive vs. negative), activation (stimulated vs. unstimulated), and dominance (in control vs. out of control) in a two- or three-dimensional space (valence and activation are the primary axes, with domination as an occasional third), which allows for various emotional experiences to be described (figure 3.1).

For instance, the emotional experience associated with the category "fear" in the context of encountering a bear in the woods would occupy a point in the intersecting dimensions of negative valence, high activation, and low dominance. Fear felt in the context of a horror movie might have less of a negative valence (some individuals may even identify with a positive valence) and definitely with neutral or higher dominance (figure 3.1). In this way, Russell (2003) argues that it can accommodate emotional categories without subscribing to the label as a fixed or universal ontology. It is thus important to understand that Russell's model of core affect is not necessarily meant to replace the social utility of emotional categories; rather, it provides a lower level scientifically sound model of affect that prioritizes

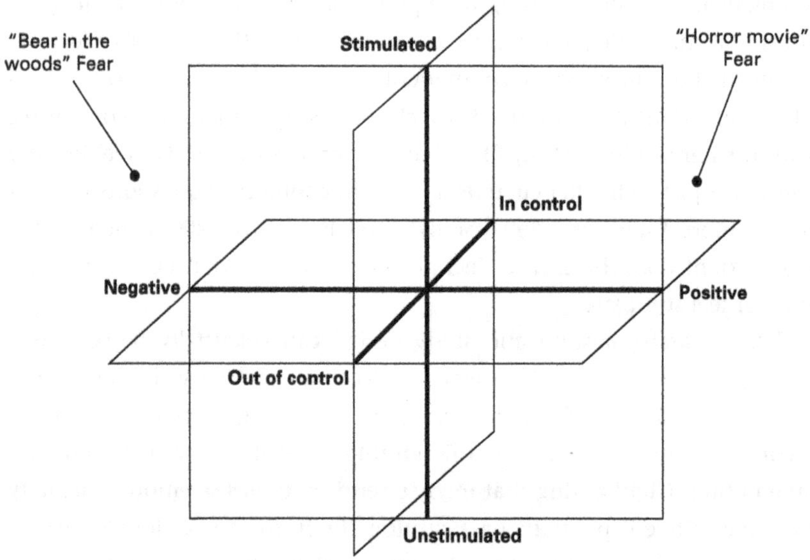

Figure 3.1
The three-dimensional core affect model proposed by Russell (2003).

transportability and irreducibility at the expense of everyday relatability. The focus is to develop *scientific tools* to describe emotional experiences writ large, of which emotional categories are a part.[13]

This background from which the dimensional theory of core affect emerges must be taken seriously in the context of Cowen et al.'s (2019) study because it changes the nuances of their hypothesis that "if categories of emotion (for example, Amusement) are psychologically constructed from more basic appraisals of core affects (Valence and Arousal), one would expect the recognition of emotion in prosody along scales such as Valence and Arousal to be better preserved across cultures than the recognition of emotion categories" (p. 371). What Cowen et al. fail to address, however, is that the core affect model was not intended as a means for the "better recognition" of emotion and was strategically and deliberately positioned against "folk theories" of emotion—that is, the categories people use to think and talk about emotion in their everyday lives—to reconcile the inconsistencies around emotional categories for the specific purpose of describing emotion within a *scientific* framework.

Cowen et al.'s (2019) hypothesis thus amounts to a straw man argument that incorrectly pits core affect against emotional categories in the context of everyday (not scientific) emotional description. That the authors find "recognition of a number of emotion categories from prosody is better preserved across cultures [by which they mean individuals living either in the US or India] than that of any of the 23 affective scales that [they] considered, including Valence and Arousal" (p. 371) is thus an expected result that is largely predictable with what discrete emotion theory and the dimensional model each stand for.[14] Not only that, but given that several of the prompts provided to the "dimensional group" also asked ostensibly ambiguous or irrelevant questions such as "To what extent does the speaker feel like his/her situation is compatible with his/her self-image?," which would inevitably require participants to have acquired supplementary information around a speaker's "self-image" based on one utterance, it is unsurprising to find that judgments made by the evaluators in the "categorical group" were more consistent.

In this way, despite their ambitious attempt, Cowen et al.'s (2019) study falls short both methodologically and theoretically in establishing the grounds for discrete emotion theory as a more robust scientific framework to other models such as core affect in the context of SER. More accurately, it represents a field-wide attempt to retroactively bound emotional ontology within the epistemological constraints of machine learning—that is, to start with machine learning as the logical foundation through which qualitative phenomena, such as emotional speech, are understood, as opposed to starting with a qualitative phenomenon and finding a method (machine learning or not) that engages with it both holistically and accurately. This is a strategic reversal from the perspective of machine learning practitioners because if machine learning is understood as the base logic through which "external" problems can be solved, nothing *can't* be distilled into a machine learning problem. Once a qualitative phenomenon is formulated into a machine learning problem, however, it should also be understood that it effectively exists as a separate "thing" tied to the epistemologies of the machine learning benchmarks, techniques, and models that define it.[15]

In the following section, I draw from GTT to attend to the processes that create this separate proxy. Specifically, I trace the actual practices of groundtruthing vocalized emotion for the purposes of building emotional speech corpora to examine not only how emotion is conceptualized in SER but also how it is elicited, recorded, indexed, and contained in SER training datasets.

Designing Emotional Proxies

Momentarily assuming that the emotional ontologies invoked in SER provide apt scientific models for classifying emotional speech, it is still important to trace how those ontologies are instantiated and materialized as ground truth. There must be material instances of vocalized emotion captured and aggregated as data upon which an SER model can train. Drawing from GTT, we can trace how such SER datasets are constructed.

Here, I examined ten highly circulated and widely cited open-source datasets used in SER applications and their accompanying research papers. Given the notorious inaccessibility of commercially developed SER applications and their underlying datasets (Boyd & Andalibi, 2023), I determined that these open-source alternatives would serve as generally representative samples of how SER datasets are constructed across both academia and the private tech industry. This is based on research that shows open-source datasets substantially inform the eventual construction of comparable privately developed commercial ones (Baio, 2022).

Some of the emotion datasets I examined include the Interactive Emotional Dyadic Motion Capture Database (IEMOCAP), a corpus that contains audiovisual data related to performances of "selected emotional scripts and also improvised hypothetical scenarios designed to elicit specific types of emotions (happiness, anger, sadness, frustration, and neutral state)" (Busso et al., 2008); the Montreal Affective Voices (MAV), a dataset comprising non-verbal affect bursts and explicitly framed as an "auditory counterpart of the Ekman faces" (Belin et al., 2008, p. 532); and the Ryerson Audio-Visual Database of Emotional Speech and Song (RAVDESS), one of the more recently published and widely circulated databases of emotional speech, among others (Livingstone & Russo, 2018). It should be noted that although some of the databases I engaged with were multimodal—for example, IEMOCAP (Busso et al., 2008), RAVDESS (Livingstone & Russo, 2018), and GEMEP (Bänziger et al., 2012)—I still included these corpora in my analysis due to a combination of their primary focus on vocal expressions of emotions (over facial) evident in the original papers and actual datasets, their frequent usage in SER applications, and their high number of citations. See table 3.2 for the full list of examined SER datasets.

The most salient theme I identified across my examination of the original papers accompanying the ten major datasets was the use of actors to perform emotional expressions to produce emotional data for the construction of ground truths. Indeed, as Bänziger et al. (2012) write in their paper accompanying the GEMEP dataset, "The procedure for *gathering* [emotional expressions] is probably the most sensitive choice to be made, one that is often determined by the theoretical assumptions of the researchers" (p. 1165, emphasis added). In practice, this procedure for "gathering" emotional expressions actually amounts to a process of "producing" them first. Here, these emotional performances produced for the purpose of representing particular emotional labels, such as "fear" or "disgust," within the datasets *stand in* as the material *proxies* for that associated emotional label, writ large.

Mulvin (2021) writes at length about the nature of proxies and their indispensable role in scientific knowledge production: "The legitimacy of an experiment, a model organism, or a statement about how the world works hinges on both a metonymic relationship (i.e., something stands in for something else) and the credibility of the stand-ins (i.e., someone believes it) to act for other phenomena that are not or could not be made present" (p. 9). For Mulvin, "Proxies operate at the hinge of ordinariness in imagining and building the world as a *controllable* space, and the locations of their use become potential sites of struggle and refusal to accept that ordinariness" (p. 19,

Table 3.2
List of examined SER datasets.

Reference	Name	Description	Method of Emotional Production
Bänziger et al. (2012)	GEMEP	7,000 audiovisual emotion portrayals; 18 emotions represented; portrayed by 10 professional actors coached by a professional director	Felt Experience Acting (FEA)
Belin et al. (2008)	MAV	90 nonverbal affect bursts; 8 emotions represented (and neutral baseline); recorded by 10 actors (5 male, 5 female)	Communication Effect Acting (CEA)
Busso et al. (2008)	IEMOCAP	151 videos of recorded dialogues including 2 speakers per video; 8 emotions represented (and neutral baseline)	FEA
Busso et al. (2016)	MSP-IMPROV	7,818 non-read (improvised or natural) speech and 620 read sentences; 4 basic emotions along with "other" option are provided as initial single-choice question, along with an additional 6 emotional categories as multi-choice question; recorded by 12 speakers (6 males, 6 females)	CEA and FEA
Cao et al. (2014)	CREMA-D	7,442 audiovisual emotion portrayals; 5 emotions along with "no emotion" option represented; portrayed by 91 actors (48 male, 43 female) coached by 1 of 2 different directors	CEA
Laukka et al. (2010)	VENEC	6,500 audio emotion portrayals; 19 different emotions represented; portrayed by 100 professional actors	CEA and FEA
Livingstone & Russo (2018)	RAVDESS	7,356 audiovisual emotion portrayals of speech and song; 6 emotions along with "neutral/calm" state represented; recorded by 24 speakers (12 males, 12 females)	CEA and FEA
Martin et al. (2006)	eNTER-FACE'05	1,166 video sequences of emotional portrayals; 6 emotions along with "neutral" state represented; recorded by 42 speakers (17 discarded), 25 speakers retained in corpus	FEA
McKeown et al. (2010)	SEMAINE	190 video sequences of interactions between a human and an operator playing a character with 4 emotionally charged personalities; 20 participants	FEA
Poria et al. (2019)	MELD	1,433 dialogues taken from the television show *Friends* (13,000 utterances); 6 emotions along with "neutral" and "non-neutral" states represented	FEA

emphasis added). The key here is control. As we will see in the methodology papers that accompany these SER datasets, the prioritization of control over ontological precision in justifying the methods used to construct emotional proxies is readily acknowledged by the dataset creators themselves. As Mulvin suggests, it is important to approach these instances of proxification with a critical orientation that queries their "ordinariness," or "taken-for-grantedness," for lack of a better phrase. Indeed, we must ask, "Who gets to choose a proxy? Who gets to use it? . . . The politics of proxy representation also mean that we have to account for how proxies are maintained and who carries out that maintenance" (p. 19).

In the context of constructing the ten major SER datasets listed above (table 3.2), the techniques used for emotional production follow two primary methods: "Communication Effect Acting (CEA)," which "requires that an expresser produces an emotion, generally specified by an emotional label—most often a label that is associated with a discrete emotion category—in such a way as to optimize the recognition of the emotion by external observers" (p. 1165); or "Felt Experience Acting (FEA)", which uses "vivid mental imagery techniques and specifically [invites] expressers to recall and relive personal past events when they felt the target emotion" (p. 1165).

Communication Effect Acting

The description of the CEA technique should be familiar. It is the same technique adopted by Dellaert et al. in their seminal paper from 1996. In order to test "statistical pattern recognition techniques to classify utterances according to their emotional content" (p. 1970), they needed to first *record* a corpus of emotional speech *produced* by actors who were given *scripts* with emotional *labels* and instructed to *exaggerate* their emotional expressions to *maximize* perceptibility. In this way, Dellaert et al. were an indispensable part of the production of the emotional proxies they were measuring: They picked the actors, wrote the scripts, chose the labels, listened to the emotions, and ultimately guided the actors' performances, until *they* felt the intended emotions were adequately audible in the recorded vocalizations. The researchers were effectively in charge of producing the very constructs they were seeking to measure.

As is evident in table 3.2, this CEA protocol for emotional proxification is also used in the construction of many widely circulated contemporary

SER datasets. One example is the MAV dataset, the self-proclaimed auditory counterpart to the Ekman faces:

> The actors were instructed to produce short emotional interjections, using the French vowel *ah* . . . and were played an auditory demonstration of the expressions that they would be asked to generate before the recording session. They had to produce vocal expressions corresponding to happiness, sadness, fear, anger, pleasure, pain, surprise, and disgust, as well as a neutral expression. Each category of vocalizations was *performed several times until our qualitative criterion was reached—that is, until the affective vocalization produced was clearly recognizable by the experimenter as the one they were asked to produce . . . Constant feedback was given to the participants during the entire session so they could improve their performance.* (Belin et al., 2008, p. 532, emphasis added)

The creators of the MAV dataset openly discuss their active participation in and meticulous guiding of the actors' emotional performances. In this way, they effectively take on the role of *emotional designers* (as opposed to data collectors) by directing and prompting their actors to vocalize in ways that adhere to their own understandings of emotion.

Semel (2022) charts a similar pattern in her ethnography with a team of researchers working on a machine listening application designed to detect signals of bipolar disorder from a caller's voice. Here, annotators were asked to "objectively" listen to and label the calls along the affective scales of valence and activation,[16] by engaging in a paradoxical mode of listening that attends only to the "sounds of the calls and not the content" (p. 12). She describes this process as "listening like a computer," in which the callers' voices are imagined as an aggregate of "disaggregate-able" modules (acoustic versus linguistic),[17] and the act of listening as a mechanistic process that can selectively filter out for those separate components. In practice, of course, Semel recounts, "There was no way to uphold the boundary between active and intentional listening (to speech form alone)" (p. 12). Indeed, despite the applaudable ethical intentions underlying the systematically implemented attempt at inattention to semantic content, the annotators'/listeners' subjectivities inevitably still colored their evaluations of the calls. In this way, the listeners' ways of listening were inextricably embedded in the underlying ground truth of this data—by "miming the imagined listening of the app" (p. 13) in the process of annotation, they also contributed to the eventual *design* of the system itself.

In the context of MAV and other SER datasets that employ the CEA protocol for producing emotions, this kind of subjective coloring is doubled: at the level of the listening researchers and the vocalizing actors. On the one hand, it positions emotion recognition as a form of verification of an underlying truth (as opposed to an individual and localized experience), and on the other, it also presumes that emotional vocalization is a universally interchangeable process not contingent on the actors. The assumptions here are that *anybody* can sound identically (or close to identically) "happy," which itself is also imagined as a universal, discrete, and abstractable emotional construct. And with the right guidance, anybody—or any*thing*—should also be able to hear it as exactly that. In Semel's (2022) words, "Annotators [and vocalizers] must format themselves into a specific kind of moral-sensory subject vis-à-vis their object of scrutiny, what Daston and Galison (2007) call the attendant 'epistemic virtue' of objectivity" (p. 15). Indeed, this means that an SER system trained on the MAV dataset, or any other dataset that follows the CEA protocol for the production of emotional expressions, ultimately takes a narrow collection of emotional vocal bursts subjectively guided by the dataset creators as its ground truth for voiced emotion.

On a more practical level, these epistemological decisions translate into urgent questions around generalizability and ethics. From a scientific perspective, how generalizable is a particular group's notion of an emotional expression beyond that context of expression? And further, from a sociopolitical standpoint, how does it normalize so-called neurotypical notions of emotional expression? Because a necessary pre-step for "hearing" and "recording" emotion in SER is to "define" and "produce" it, the process inherently becomes a normative one that is embedded with assumptions held by both producer and recorder.

As sound studies scholar, Nina Eidsheim (2018) argues, "the assumption that we can know sound, and that the meaning we infer from it is stable (and indeed essential), allows for the *projection of beliefs about people onto the sound*" (p. 49). By "listening to listening"—that is, examining *how* one listens, or in this case, examining *how* the voicescape an SER system listens to/through is constructed—it becomes possible to recognize that the self-proclaimed "universal" emotions collected for MAV using CEA are actually a collection of narrowly defined and guided emotional performances. Here, not only their underlying objectivity but also their practical efficacy as measurable constructs are contingent on a host of epistemologically questionable assumptions.

Felt Experience Acting

Other datasets such as the Interactive Emotional Dyadic Motion Capture Database (IEMOCAP; Busso et al., 2008) and the GEneva Multimodal Emotion Portrayals (GEMEP; Bänziger et al., 2012) try to circumvent the perceived "inauthenticity" of performed emotional expressions in CEA by instructing actors to subscribe to the aforementioned FEA technique. This includes the use of hypothetical scenarios, interactions, monologues, and improvisations to "increase the authenticity" of the emotional expression so as to argue that the emotion is "elicited" as opposed to "performed." Even here, however, the successful production of emotional voice data is contingent on close collaboration between the researchers and actors. STS researcher Isak Engdahl (2024) refers to this "hands-on stewarding" (p. 1) in the production of data as "alignment work." This kind of intervention and inevitable calibration toward particular objectives and formats "destabilizes the notion of the dataset as 'unscripted' or captured 'in the wild', suggesting instead that data subjects performed activities in accordance with their alignment" (p. 10).

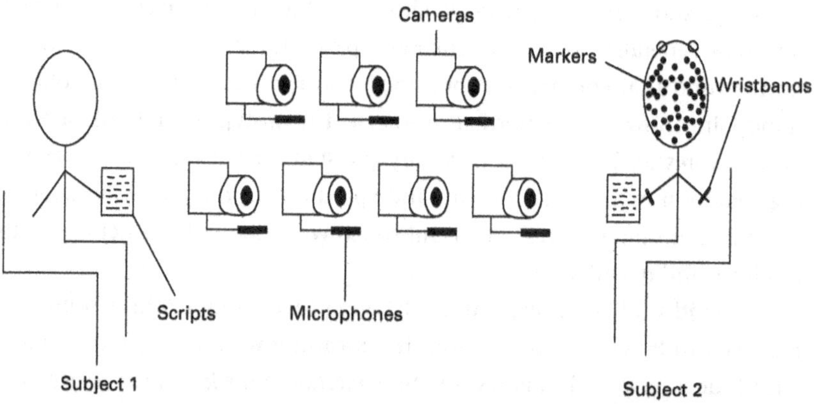

VICON motion capture system with 8 cameras. The subject with the markers sat in the middle of the room, with the cameras directed to him/her. The subject without the markers sat outside the field of view of the VICON cameras, facing the subject with markers.

Figure 3.2

A labeled redrawing of a photograph on page 10 of Busso et al.'s (2008) article titled "IEMOCAP: Interactive Emotional Dyadic Motion Capture Database." It depicts the setup that Busso et al. used to record actors expressing emotions using the FEA method. The original caption accompanying the photograph from Busso et al. (2008) is also included with the redrawing.

In this example depicted in figure 3.2, the intricate recording setup alone, which includes eight cameras, two wristbands, a headband, and fifty-three markers physically attached to the face, should further call into question the apparent "authenticity" or "naturalness" that the FEA protocol is meant to introduce. Indeed, existing empirical literature that examines the effects of video and audio recording on therapist–patient interactions in psychotherapy sessions has shown that the presence/awareness of much less intrusive recording devices still affects how patients and therapists express themselves, such as patients making more positive self-references (Roberts & Renzaglia, 1965) and/or showing higher levels of inhibition (Tanney & Gelso, 1972), as well as therapists exhibiting increased anxiety (Yenawine & Arbuckle, 1971) and/or increased negative self-ratings of performance (Niland et al., 1971). Such effects of recording are not exclusive to FEA, of course, and equally influence the CEA method. But it does call into question the purported "naturalness" of the FEA setting, which is central to the argument that the FEA protocol allows for increased levels of authenticity to be elicited in the actors' emotional expressions.

In addition to the confounding effects of recording on FEA's claim to authenticity, it's also important to note that because all implementations of FEA qualify as human-subjects research, facilitators must explain in their Institutional Review Board (IRB) protocol how they intend to protect participants from experiencing any kind of psychological discomfort or harm. It's unclear how emotional experiences that *require* psychological discomfort such as fear, anxiety, or sadness are *ethically* elicited. Indeed, there are no direct mentions of how this conundrum was addressed in any of the original papers accompanying the major datasets I examined. Instead, they all outsource the validity of their methods by emphasizing a steadfast belief in the ability of the participants (some of whom were expressly not actors) to effectively situate themselves in the appropriate mental state to express the target emotions. In the absence of any mentions of how this elicitation process was framed for IRB approval, it can only be deduced that it ultimately had to be redefined in the proposal as one in which the emotional expressions related to psychological discomfort were not really "felt," or felt in a way that did not adequately capture what one may consider the authentic psychological discomforts related to disconcerting emotional experiences. This puts into question the efficacy of FEA in addressing the methodological limitations of CEA: How emotionally authentic is a non-psychologically disconcerting anger or sadness, and how is that different from the anger or sadness produced using CEA?

In relation to this, Bänziger et al. (2012) dedicate a section in their paper accompanying GEMEP titled "In Defense of Felt Experience Enacted Emotion Expressions." Here, they accurately elaborate on the precarity and mediated nature of "naturally occurring EEs [emotional expressions] captured in everyday life," stating that there are "extraordinary ethical and practical difficulties of recording private, largely unpredictable, and fleeting episodes for a sufficient number of individuals" as well as methodological difficulties related to "sampling a large number of different emotions per person that are comparable in nature and context across individuals" (p. 1174). Not only this, but the authors themselves also emphasize the inherently mediated nature of emotion referencing scientific literature that shows "emotional expression is rarely free of posing" (p. 1174). Critiquing studies that used emotional expressions recorded from television shows as ground truth for "emotions-in-the-wild" (e.g., Poria et al., 2019), the authors state that this mediation effect is "even worse in EEs [emotional expressions] recorded form the media (as in reality or game shows), given that in addition to ordinary display rules and regulation attempts, there may be systematic bias resulting from coaching of participants by the organizers or audience reaction" (Bänziger et al., 2012, p. 1174).

What is astonishing about this relatively thorough and accurate description of the tenuousness of recording natural (i.e., "authentic") emotions given by the authors, however, is its juxtaposition against a purely utilitarian justification for FEA that emphasizes *experimental control* over ontological precision. Although the recruiting of professional actors and recording them in controlled settings does indeed allow for systematic experimentation, it does not circumvent any of the fundamental complications related to the stability, generalizability, and authenticity of emotional expression that the authors themselves critique. Even if one accepts the emotional authenticity of an actor's "attempt to enact as faithfully as possible the recalled or imagined emotional experience in the expression shown" (p. 1174), this does not adequately address the fact that the production of emotion using FEA always creates an *instance* of vocalized emotion tied to the context of production.[18]

Put differently, a performed emotion is still arguably an emotional experience, but it is not an emotional event/expression/experience that is generalizable beyond that instance. A woman reciting a sentence accompanied by the label "joy" in a way that both she and the study facilitators perceive to be joyful may indeed be an instance of joy, but it does not represent a generalizable (neither across- nor within-subject) representation of joy. This is true regardless of whether the emotional expression is the product of a script or an "authentic" elicitation. An adequate engagement with the multiplicity of emotional ontology and existing literature on emotion science—which would require acknowledging that there is a glaring lack of scientific evidence suggesting a universal experience of emotion exists, and that the attempt to argue such a claim materializes as a form of emotional regulation that normalizes particular emotional experiences over others—shows that the limitations of SER manifest not only in the process of ground-truthing but also in the underlying premise itself.[19]

Indeed, drawing from emotion literature in psychology, existing critical scholarship on AI emotion recognition has accurately pointed out that "external" expressions of emotion comprise only a part of the broader emotional experience, and are more accurately and consistently reflective of social motivations, while providing little evidence of "interior" mental states (Boyd & Andalibi, 2023; Stark & Hoey, 2021). Not only that, but the unavoidable effect of culture, broadly construed, as a powerful mediator in the way emotions are expressed *and perceived* (Gendron et al., 2015; Leshin et al., 2024; Russell, 1991) further sheds light on the limitations of a generalizable emotion recognition system.

Leshin et al. (2024), for instance, "investigate[d] the impact of cultural influence and emotion concept accessibility on emotion perception" (p. 1) by using functional magnetic resonance imaging technology "to access functional connectivity between brain regions during emotion perception" (p. 1) across two distinct cultural groups:

1. Chinese participants who were "born and raised in provinces of mainland China excluding areas with significant Western influence, such as Hong Kong, Macau, and Taiwan" (p. 3) and had resided in America "for an average of less than 20 months" (p. 4).
2. "Monolingual English-speaking non-Hispanic White American" participants who had "denied ever residing outside of the United States" (p. 3).

Based on their results, which indicated significant differences in neural responses between Chinese and White American participants to emotional labels such as "disgust" (and stimuli meant to represent "disgust"), the researchers suggest that "both culture and access to emotion category words may influence how the brain represents emotional facial behaviors during emotion perception" (p. 6). Although this particular study engages with the emotion perception of *facial* behaviors, the focus should be on the empirical measurement of the nontrivial effect of culture on subjective emotional perception, writ large. The question is thus about not only who is guiding the emotional expression but also from *whose perspective* the emotion is being evaluated. Subjective perception/listening is present throughout the development pipeline of machine listening SER systems from conceptualization, production, and evaluation.

Examining both the theoretical framework through which emotion is defined (discrete emotion theory) as well as the methodological choices for producing emotion at the level of ground truth (CEA and FEA) sheds light on the weight of utilitarian fit as a mode of decision-making in conceptualizing and building machine listening SER systems. This prioritization of functional proxies over scientific depth in machine learning is not unique to SER, but it does have particularly concerning implications as it relates to the implementation of SER across key social infrastructures.

In the final section, I build on perspectives from disability scholars to trace the sociopolitical stakes of SER. Specifically, I examine its role as a regulatory surveillance system that applies speech emotion recognition to the context of optimizing labor productivity in call centers employed by American businesses.

Disabl*ing* Technologies

Scholars working at the intersection of disability studies and technology (e.g., Alper, 2018; Mills, 2010, 2011, 2015; Nagy, 2022; Whitaker et al., 2019) have long pointed out that the "concept of 'normal,' as well as the tools and techniques for enforcing normalcy, have historically constructed the disabled body and mind as deviant and problematic" (Whitaker et al., 2019, p. 3). In the developmental history of machine learning, however, which operates under the epistemological domain of statistics and therefore necessitates normative models in their successful operation, this understanding of disability has often served as a "charismatic" use case (Ames,

2019) for the development and justification for "assistive" technologies,[20] as well as been unsettlingly used as a metaphor for the deficient computer.

"Helping Autistic People" (p. 89), for instance, is one of the section titles and listed applications of affective computing in Picard's 1995 text. Here, she states, "One of the hallmarks of [autism] is difficulty with emotions—recognizing the meanings of other people's emotions, suitably expressing emotions, and having empathy" (p. 89). She goes on to elaborate that "computers are like autistic people—particularly like autistic 'idiot savants,' an unfortunate term that has been used to describe people who have unusually gifted abilities in certain areas—such as rapid computation of large numbers, memorizing phone listings, and precise memory of huge sets of facts and trivia, but who lack the forms of common sense and emotional intelligence that most people acquire effortlessly . . . Unaffective computers are similarly *handicapped*" (p. 90, emphasis added). Leveraging both a metaphorical alignment between mental ability and computation, as well as presenting automated emotion recognition as an educational tool and "assistive" technology for autistic individuals, Picard effectively casted automated emotion recognition in a prosocial light. This allowed her to garner substantial grant support from the National Science Foundation, the National Institutes of Health, and the National Institute of Mental Health to support her research into affective computing (Nagy, 2022, p. 5).

The problem with this metaphor, understood as either the computational model of autism—that is, likening autistic individuals to "computers running a poorly implemented software architecture, one that, insofar as empathy is concerned, suffer[s] from an internal signal processing disorder that ha[s] concomitant external effects on social signal processing" (p. 7)—or the autistic model of computers—that is, "handicapped" machines that fail to interpret and respond to users' emotions—is that it adopts a "medical model" of disability that understands autism as a *deficit* of the individual. Relying on biomedical standards of "normal" bodies, such a view pathologizes conditions outside of these norms as impaired and thus in need of augmentation or correction (Siebers, 2008). This is in opposition to a social model of disability, which understands it as a *relational* product contingent on environments and attitudes, wherein "the locus of intervention [is] not at the level of the individual—with efforts that would attempt to "fix" disabled people—but at the level of social justice" (Whitaker et al., 2019, p. 4).

Conceptualized and developed through its medical framing as a "socio-emotional prosthetic" for autistic individuals (Nagy, 2022, p. 8), AI emotion recognition thus effectively abstracts autism away from the diversely experienced condition that mediates autistic individuals' lives into a standardized and fixed pathology that can be used to denote empathy "impairment" in both humans and machines. In recounting the development of El Kaliouby and Robinson's (2005) "emotional hearing aid," a facial affect prosthetic for children with Asperger's syndrome, Nagy observes that their adoption of a medical model of autism "prompted the system's users to shape their emotional expressions to meet the expressions of non-autistic social others" (p. 10). Based on real-time video assessment of detected faces, the emotional hearing aid instructed users to "apologize or explain for what the system register[ed] as a neurotypical interlocutor's disgust or confusion [which] place[d] the burden of rectifying an apparent 'conversational impediment' in the interests of producing a 'productive state' entirely on side of the autistic individual" (p. 10). In this way, it served as a disciplinary and normative technology that trained autistic users toward "becoming docile and conciliatory conversation partners" (p. 10) to their non-autistic counterparts.

This historical relationship between affective computing and autism research is a characteristic instance of what Mills (2010) calls "assistive pretext" (p. 39), wherein technologies initially developed to address disability-related use cases are reconfigured and made relevant to the general public. The translation from socio-emotional prosthetic to commercial emotion AI, which effectively locates emotion recognition within the extractive domain of surveillance capitalism (Zuboff, 2019), sheds light on a "productive parallelism between the ways that autism mitigation programs view autistic individuals and the ways that surveillance capitalism views platform users. Both view the subjects in their purview as collections of manipulatable, infra-individual behaviors instead of as independent actors capable of psychological depth" (Nagy, 2022, p. 12). Disability, understood as pathology, and medical frames of "disability correction" thus comprise the sociotechnical interface through which commercial AI emotion recognition can be understood to regulate its subjects. It effectively constructs a system in which subjects are required to make their emotional experiences legible to the systems that detect it, thereby flipping the directionality of power from one that, in principle, should be able to "recognize" emotional expressions in all of their multiplicity to one that, in practice, "constrains" them to a particular set of normative and productive behaviors. This "reverse Turing-test" situation (Nakamura, 2019), in which humans must prove their humanity to AI systems,[21] is also highly representative of the way that SER is currently deployed in commercial contexts.

SER in Call Center Optimization

Cogito is a software company that applies SER to the context of call center optimization. In monitoring conversations between call center operators and customers, the company "extracts and analyzes over 200 acoustic and voice signals in milliseconds to give . . . agents cues on how to adjust their behavior and surface the best recommendations" ("Cogito," n.d.). It is promoted as a human-centered AI tool that improves customer experience (CX), in which "CX measured on every call delivers the data for personalized coaching and development plans, essential to growing and retaining your employees" ("Emotion AI," n.d.). According to its website, Cogito is used by "3 of the top 5 US telecom technology companies," "2 of the top 5 US cable providers," "4 of the top 5 national health insurers," and "3 of the top 5 pharmacy benefits managers," of which eight are Fortune 25 brands ("Emotion AI," n.d.). Presuming this to be true, this means that Cogito's products span multiple industries and have largely become an embedded part of the CX in call centers used by American companies.

There are important parallels to consider in the way Cogito frames its use case as helping call center operators through SER with the aforementioned histories of affective computing and autism. A 2018 article in *Wired* references Cogito's chief operating officer Tracy Dudek stating that positive emotional evaluations of agents by its SER system can boost agents' chances of performance-related bonuses (Simonite, 2018). The potential flipside of this, of course, is a performance-related *penalty* related to behaviors that might be categorized as a "negative emotional evaluation,"[22] a feature also observed across other AI emotion recognition systems used for worker evaluations described in various patent applications (Boyd & Andalibi, 2023). Dudek's statement reveals an explicit link between emotional regulation and worker productivity facilitated through a surveillance tool that constantly monitors operators' conversations. Understood as assessing a modular assemblage of adjustable behaviors as opposed to independent individuals with the capacity to experience a range of emotions, Cogito's SER effectively serves as a disciplinary mechanism that systematically fastens worker compensation to

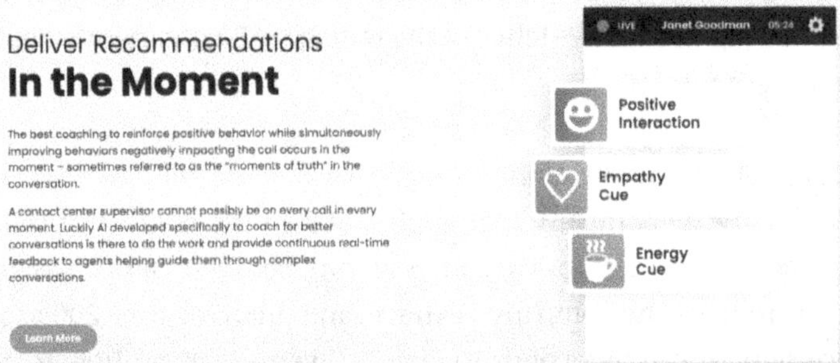

Figure 3.3
A screenshot taken in March 2023 of Cogito's emotion detection system for call center workers.

emotional regulation so as to enforce workers to self-regulate their behaviors to be positively read by Cogito's SER system (figure 3.3).

Roemmich et al. (2023) describe the role of AI emotion recognition in determining emotional measures of desirability for subjected workers as the "*affective commodification* of a candidate's *affective value*" (p. 1). As I have elaborated upon throughout this chapter and the previous one, however, an SER system, like any other machine learning technology, is highly contingent on its ground truth, wherein the ground truth does not necessarily denote any kind of "real" qualitative truth but rather reflects the ontological alignments and epistemological assumptions of its creators. In other words, the ground truth for "positive evaluations of emotion" in Cogito's SER system is a construct designed by Cogito itself. The company's rhetoric of human-centered design and improving self-satisfaction of agents through empathy coaching thus denotes an arguably unproductive self-contained loop in which operators have the additional labor of making themselves legible to the data structures now entwined in their performance reviews and, as an extension, livelihoods.

There are also important practical implications of applying SER to CX on the side of influencing customer behavior. In the "Ubiquitous Technologies for Emotion Recognition" special issue of *Applied Sciences*, Bojanic et al. (2020) propose a system in which call centers can rank the urgency of calls based on SER, "giving greater priority to calls featuring emotions such as fear, anger, and sadness, and less priority to calls featuring neutral speech and happiness" (p. 1). Juxtaposing research that documents the psychological strains of abusive behavior by "angry" callers on call center employees (e.g., Rohrmann et al., 2011) alongside the explicit use case invoked by companies such as Cogito—to "Reduce Employee Churn" ("Reduce Employee Churn," n.d.)—a proposition to prioritize calls that are audibly "angry" (i.e., to fit the sonic profile of a prototypically "angry" voice that the developers have agreed is what constitutes "anger") will presumably have an even greater negative effect on the mental health of call center operators. Indeed, it sends an obvious message for how to "game" the system: sound angrier. Not only that, but by focusing specifically on the "empathy coaching" of agents, thus disciplining them to better regulate both their own and callers' emotions, the application of SER to call centers ultimately strips power away from the operators.

This is similar to how El Kaliouby and Robinson's (2005) emotional hearing aid was sold as a tool to help autistic individuals, but in reality, resulted in assimilating them to ableist standards of conversational productivity. By prompting autistic users to rectify "neurodivergent" behaviors, the emotional hearing aid placed neurotypical behaviors as the singular mode of legible—that is, acceptable—emotional expression, without ever fully engaging the possibility of a wider spectrum of emotional productivity. Indeed, as a concept that was *invented* in the nineteenth century directly alongside the notion of the wage economy (Rose, 2017) "when employers began to use preemployment screenings to eliminate people deemed inefficient, nonproductive, and likely to require extra help and support" (Whitaker et al., 2019, p. 17), disability cannot be understood separately from the notion of worker productivity. In this history, the lack of ability to conform to particular notions of productivity effectively rendered an individual "disabled," thereby disqualifying the person from participating in the workplace and earning a living.

Cogito's use of SER as a mode of worker surveillance continues this history of ableist workplace politics in which its subjects are located on an expanded spectrum of emotional suspects that require constant monitoring and validating. As part of a scientific lineage that is intimately linked with adopting a medical (i.e., deficit) model of autism, and then applying that model as the default framework for understanding its users beyond the original context of disability, Cogito's SER can also be interpreted as a disabl*ing* technology that transposes the systemic oppression of disabled individuals on to its users. To be clear, this is by no means to argue that the systemic subjugation of disabled people should be contained, but rather the opposite. It is to problematize the medical model of disability altogether so as to argue that such ways of thinking ultimately make such oppressive structures more widespread.

* * *

The datasets I examined in this chapter comprise the machine voicescapes through which SER systems understand and "hear" emotion. If the voicescape for an embodied human listener describes the broader vocal ecologies (Ceraso, 2022) and imaginaries (Eidsheim & Meizel, 2019) that mediate the ways in which we imbue subjective meaning to certain types of voices, the training data that underlie machine learning systems serve the same purpose for the "listening machine." These training datasets, however, are intimately linked to the human subjects that participate in their making.

Applying GTT to the ten SER datasets and methodology papers examined in this chapter made this apparent. Here, dataset creators adopted the role of *designing* vocalized emotion, all based on their own assumptions of what emotion is, what it sounds like, and how it should be measured. As such, the epistemic constraints of machine learning, in which the knowledge produced from a machine learning model is inextricably tied to its ground truth reference, combined with the highly contested nature of this particular domain—the fact that no one can agree on what emotion is or how to scientifically study it—renders the foundational premise of SER more of an aspirational fantasy for the AI community than a reality. If anything, the attempt to build an emotion-detecting "listening machine" ultimately results in the normalization of quasi-functional data proxies for emotion that fail to encapsulate either its ontological breadth or depth, while nonetheless inheriting its cultural significance and social capital.

We can see the material effects of such data proxies through Cogito's application of SER to call center optimization, which ultimately produces a disciplinary system that divests power from call center agents and consolidates it within its creators. This technological displacement of power, of course, is a tried-and-true critique of the broader power relations that mediate technology creators and users, writ large. Indeed, my own situating of Cogito within a longer history of affective computing research demonstrates the normative and disciplinary functions of emotion recognition technologies that have come long before Cogito's SER.

My specific intervention, however, lies in the reinforcement of this critique with GTT, through my empirical engagement with the grounded practices, techniques, methodologies, and decisions that shed light on exactly *how* these proxies are constructed and what they show about the tenuousness of the promises of these technologies. This intervention is a broader push toward interdisciplinary collaborations between "technical" and "sociotechnical" communities (for lack of a better semantic distinction), as well as a more focused and urgent call to reevaluate the limitations underlying the premise of SER.

Although the application of SER to call centers currently represents the most widely observed commercial use case led by technology companies such as Cogito, Nuance, and Behavioral Signals, among others, there is a wide variety of other emerging applications. Amazon's emotion tracking wearable device, "Halo," for instance, monitors and analyzes a user's tone of voice to make wearers more aware of how they sound to others (MyHealthyApple, 2022). Voicesense offers a suite of voice analytics products, one of which apparently measures emotional states from the voice to screen and monitor mental health risks, such as depression ("Mental Health and Wellness," 2022). Affectiva (now a subsidiary of Smart Eye), the facial emotion recognition company founded by none other than El Kaliouby and Picard, also promotes the importance of SER as it seeks to develop *multimodal* AI emotion recognition systems in automobiles ("Introducing Affectiva's Emotion Recognition Through Speech," 2017). Given the comparative slowdown in facial emotion recognition development with Microsoft's decision to retire its emotion recognition technologies from its Azure Face facial recognition services (Bird, 2022), it will also be important to bring attention to the dubious scientific and ethical implications of SER before it becomes a more ubiquitously adopted technology.

Interlude 2 Beyond Listening, Toward Voicing

September 21, 2023: Today Was the *Her* Class

As part of their weekly homework for the "AI and Society" undergraduate seminar I teach, I asked my students to watch the Spike Jonze film from 2013. It's a film almost never left out of discussions around AI futures and AGI, and for good reason: despite its pensive ending, its characterization of Samantha, the "OS" (operating system) voiced by Scarlett Johansson, and her relationship with Theodore, the human protagonist portrayed by Joaquin Phoenix, and Samantha's boyfriend, unnervingly resembles the AI futures that have come to dominate our cultural imagination since the advent of technologies such as ChatGPT.

My intention for assigning the film was to problematize this tendency to anthropomorphize AI. And yet, it was admittedly hard to resist the undeniable presence that Samantha—despite never having any true visual representation, except as Theodore's phone, which arguably *minimizes* her presence—commanded throughout the film. Indeed, our class discussion around *Her* primarily revolved around the different ways that Samantha both felt and didn't feel *human*. Students remarked: "Theodore was so right to ask why Samantha sighs. It feels performative!" "Is Samantha really feeling an orgasm?" "How does she feel pain?" And yet, others also commented: "I almost forgot that we never actually see Scarjo[1] in the film." "When they go on the double date, it totally feels like she's just there with them." "When she says she's hurt, it feels convincing, even though I don't understand how." Discussions around Samantha's *in*humanity returned to her lack of a body; discussions around her humanity—to the fullness of her voice.

Voices in machines represent this paradox of a disembodied voice. But AI voices, of course, are not the first time that "the voice without a body" has been the subject of intellectual inquiry. Film theorist Michel Chion (1982/1999) has written extensively about the "acousmatic" voice in filmmaking: "a sound that is heard without its cause or source being seen" (p. 18). Describing this disembodied voice as possessing "a sort of virginity" (p. 23), Chion says that at the moment in which the body that emits the voice is shown on screen, the acousmatic voice "re-enters the realm of human beings" (p. 23), thereby losing its apparently divine properties. Its power lies in its disembodied state.

Chion's (1982/1999) connection between the disembodied voice and an apparently innocent godliness presents an allegory through which we might also frame the ways in which the "datafied voice," in all of its different iterations, is displaced from the body that emits it and re-situated as a sanitized proxy for machine learning applications. Indeed, this is how abstraction operates. By removing the subjectivities that are inherently a part of a phenomenon's construction and essence, and then reconstructing the phenomenon as an objective entity of its own, abstractions acquire the powers of "ubiquity, panopticism, omniscience, and omnipotence" (p. 24). This allows them to travel flexibly as "norms" as opposed to being bound in specificity. Despite Samantha's simultaneously prophetic and awkward acknowledgment of this power of disembodiment during the picnic with Theodore's work colleagues—

> You know, I actually used to be so worried about not having a body, but now I truly love it. I'm growing in a way that I couldn't if I had a physical form. I mean, I'm not limited. I can be anywhere and everywhere simultaneously. I'm not tethered to time and space in the way that I would be if I was stuck inside a body that's inevitably going to die. (*Her*, dir. Jonze, 2013)

—her all-too-human voice ironically conjures a spectral body through which the audience is invited to situate her humanity.

Thinkers from an eclectic range of disciplines have reflected on this relationship between the voice, body, and human condition. Feminist philosopher Adriana Cavarero (2005), for instance, calls attention to what she refers to as a vocal phenomenology of uniqueness, "an ontology that concerns the incarnate singularity of every existence insofar as she or he manifests her- or himself vocally" (p. 7). Cavarero's emphasis on the materiality and corporeality of voice—or "throat of flesh" (p. 234)—is evident in her opening reference to Italo Calvino's story of the listening king: "A voice means this: there is a living person, throat, chest, feelings, who sends into the air this voice, different from all other voices" (p. 1).

Cavarero's position is indicative of an earlier moment in the emerging stages of "voice studies" as a field, when theorists of voice were understandably preoccupied with distancing their understandings of voice from the perspective provided by the related but significantly more developed field of linguistics. By focusing on the materiality of voice, Cavarero explicitly dissociated the study of voice from the study of language, in which voice is framed only as the carrier of semantic meaning. Indeed, in linguistics, the distinctiveness of voice is intentionally suppressed for it to serve its primary purpose as a communicative medium: The voice "makes the utterance possible, but it disappears in it, it goes up in smoke in the meaning being produced" (Dolar, 2006, p. 15). As philosopher and cultural theorist Mladen Dolar elaborates, however, the *timbral* voice is that which "*does not contribute to making sense.* It is the material element recalcitrant to meaning, and if we speak in order to say something, then the voice is precisely that which cannot be said" (p. 15).

In this sense, Cavarero is justified in resounding the timbral voice through an invocation of the body. Indeed, as musicologist Laura Turnbridge (2016) writes in "Scarlett Johansson's Body and the Materialization of Voice," Johansson's voice "possesses a kind of tactility associated with [Barthes's] 'grain of the voice' ('it crackles, it caresses, it grates, it cuts, it comes') [1981]" (p. 144). She calls this the "haptic voice," which "encourages a conception of voice that may sound as if it is based in the body, but need not be; the voice can be technologically mediated, even constructed, but feel no less real" (p. 145). The haptic voice thus provokes a reflection on the relations of the voice and the human condition between which the body sits as a bridge. Through vocal performance—the sighs, moans, and huskiness—the haptic voice forges a body where there is none.

Such ruminations on the relations between the body and voice bring forth three foundational questions that will guide the chapters to follow: What is a voice without the body and the subjectivities through which it is sounded? What is a voice synthesized from the sounds uttered through multiple bodies? And why might such a voice be useful for the different actors interested in engineering these disembodied vocal chimeras into existence?

Here, Chion's (1982/1999) deliberations around the "acousmatic voice" as well as Turnbridge's (2016) framework of the "haptic voice" together present an alluring tension with which to reflect on such relationships between the voice and body. Whereas the acousmatic voice describes the godlike omnipresence of disembodiment, wherein the subjective and the corporeal are abstracted away, the haptic voice, through various techniques of vocal performance such as sighing and moaning that conjure sonic vestiges of the material body, resituates the acousmatic voice in a spectral body. Connor (2000) refers to this as the "vocalic body"—a "surrogate or secondary body, a projection of a new way of having or being a body, formed and sustained out of the autonomous operations of the voice" (p. 35). The result, as we hear in Scarlett Johansson's voicing of Samantha, is an enticing chimera that is simultaneously multiple, singular, and absent.

In this way, the "acousmatic voice," the "haptic voice," and the "vocalic body" are all frameworks that parse the ontological hierarchies and relations between voice and body. As Connor (2000) writes, "Voices are produced by bodies: but can also themselves produce bodies" (p. 35). Writing against Altman's (1992) "sound hermeneutic," which asserts that the acousmatic voice must still be associated with a plausible material body, Connor argues that this suggests "the voice is subordinate to the body, when in fact the opposite is experientially the case; it is the voice which seems to colour and model its container" (p. 35). As I briefly examine below, both appear to be true in the various forms of embodied and AI-enabled "racial passing" (Burma, 1946).

The Vocalic Body and Passing

Studies have documented how Black individuals often adapt "White ways of speaking" on the phone to avoid discrimination—such as when inquiring about loans for housing (e.g., Henderson, 2001). Viewed through Connor's (2000) frame, it can be said that such individuals leverage a temporary "vocalic body" erased of its Blackness. In so doing, however, these individuals also invoke the "sound hermeneutic" through the *performance* of a plausible body that presents itself literally as a more "acceptably White" one. Here, this act of vocal passing can be interpreted as a response to counter historically documented racialized listening practices of auditory redlining that limited Black bodies from entering White communities (Horwitz, 1999).[2] Vocal performance can thus not only recolor the surrogate vocalic body but also reinforce the indispensability of the racialized body, albeit a different, temporary, and sonically forged one. It "conjures for itself a different kind of body; an imaginary body which may contradict, compete with, replace, or even reshape the actual, visible body of the speaker" (Connor, 2000, p. 36), but only to the extent that it temporarily replaces it with another plausible or coherent one.

The Sound Hermeneutic and Failing to Pass

In 1992, communication researcher Donald Rubin published an experimental study in which undergraduate students were asked to score the "comprehensibility" and perceived "intelligence level" of a lecture recorded in Standard American English (Rubin, 1992). Here, one group listened to the lecture accompanied by a photo of a White person, and another group listened to the exact same lecture but accompanied by a photo of an Asian person. The study found that the students in the listening group presented with the photo of an Asian person consistently gave lower ratings for both comprehensibility and perceived intelligence. The results of this study would suggest that it is not necessarily the *White-sounding voice* but rather the *imagined White body* from which the voice emanates that mediated how the students "heard" the lecture. Indeed, vocalic bodies can literally change what one "hears" in the same voice and most definitely any meaning inferred from it.

Passing with AI

"Real-Time Accent Correction AI" by Data Monsters is an AI translation tool that changes the voices of call center agents to "transform non-native speech to native speech" to "facilitate conversations between call centers and clients from different regions" (Data Monsters, 2023). Here, vocal passing is made possible not through the body but rather through computational processes. Justifications for this technology often invoke research that suggests the difference between "successful" and "unsuccessful" call center agents comes down to "communication issues" (Friginal, 2009). This might be attributed to the fact that call center workers are usually from countries of the Global South such as the Philippines or India, where English may not be their first language. Data Monsters thus situates its locus of intervention at the level of altering the call center agents' voices. This is often justified as a "win-win" situation for the worker and the caller: The caller is presented with a worker whose English is "unmarked" and therefore presumably easier to understand, and the worker is protected by the sonic shield of a White-sounding voice.

Viewed through the results of Rubin's (1992) experiment, however, it might also be said that a key element enabling this so-called win-win situation is the production of temporary White vocalic *bodies* onto which the AI-edited voices are projected.

* * *

Both Altman's (1992) sound hermeneutic and Connor's (2000) vocalic body are concepts that have, to varying degrees, engaged with the representation of the disembodied voice in audiovisual media,[3] but the examples above suggest how they can shed light on the ontological tensions between voice and body *beyond* media representations. Indeed, although these frameworks differ in their hierarchical conceptualizations of voice and body—which colors which?—they are alike insofar as their dependence on the inseverable link between the two. The body produces a voice, which is why the voice can also conjure a body.

The final chapters of this book center machines that produce voices: "speaking machines." As we will read, their production is intimately tied not only to the conjuration of vocalic bodies that are constructed alongside such machine voices but also to *human bodies* that must come together to voice, listen to, and curate the data that underlie these synthetic voices. Here, chapter 4 delineates my interviews with voice actors who are indispensable to the production of synthetic AI voices. Chapter 5 ventures into the business of AI voice generation and the economic standards that mediate its development.

III Speaking Machines

4 Voice Engines

The ChatGPT Moment

OpenAI's release of Chat Generative Pre-trained Transformer, more commonly referred to as ChatGPT, on November 30, 2022, was revolutionary, to say the least. Despite CEO Sam Altman's own admission that it's a "horrible product" (D'Cruze, 2023)—citing its "frequent error messages, simplistic design, and capacity issues"—ChatGPT's "ability" to "converse" via text in a seemingly "natural" way marked an unprecedented moment of technological spectacle and public awe. Indeed, it now felt as if artificial intelligence *had a voice.*

According to Forbes, the introduction of ChatGPT was enough of a groundbreaking moment that the company labeled 2022 as "the year that AI took center stage" (Bajarin, 2022). This also coincided with the marked transition away from the use of the term "machine learning" (which theoretically is a subset of the broader concept of AI but, as mentioned in chapter 2, comprises the vast majority of how existing AI systems, including ChatGPT, are developed) and a reversion toward the more ambiguous but exciting term "AI."

Why was the introduction of a chatbot the watershed moment for artificial intelligence? Arguably, Google Deepmind's monumental development of AlphaFold, a machine learning model released only a year before ChatGPT that can "predict the 3D structure of a protein just from its 1D amino acid sequence" (Jumper et al., 2021), was a significantly more substantive development in the application of AI to real-world problems. It addressed one of the longest standing challenges of the scientific community, revolutionizing the biotechnology industry by enabling chemists to discover drugs at a significantly more efficient rate. Other machine learning systems such as facial recognition software, online visual search, social media newsfeeds, content recommendation, and all of the machine listening technologies examined thus far in this book are existing AI applications that have drastically reconstituted how society is organized. And yet, it was OpenAI's chatbot that put AI on center stage.

On September 25, 2023, OpenAI doubled down on its infamous chatbot, publishing a blog post titled "ChatGPT can now see, hear, and speak" (OpenAI, 2023). The company wrote: "The new voice capability is powered by a new text-to-speech model, capable of generating human-like audio from just text and a few seconds of sample speech. We collaborated with professional voice actors to create each of the voices. We also use Whisper, our open-source speech recognition system to transcribe your spoken words into text" (OpenAI, 2023). Ten months after its initial release, and with the help of human voice actors and machine learning practitioners, ChatGPT *really* "had a voice."

"The ChatGPT moment" represents a unique opportunity to reflect on the centrality of voice—which has fueled our cultural imagination of AI for decades but has rarely been deeply investigated from a scholarly perspective[1]—to our understanding of artificial intelligence. It's important to acknowledge, however, that ChatGPT wasn't the first AI system to "speak." Apple's Siri and Amazon's Alexa are both common household products that preceded ChatGPT in exhibiting so-called vocal features. In fact, the original "chatterbot," ELIZA, developed by Joseph Weizenbaum almost six decades earlier in 1966, already demonstrated the early capabilities of NLP in "conversing" with humans via text. Any of these systems can be claimed to have marked the moment that AI was "given a voice." But the spectacular contextual flexibility of ChatGPT—in comparison to the considerably bounded vocal and linguistic capabilities of preceding systems—"felt different," as one student in my "AI and Society" undergraduate seminar remarked.

This moment of intelligible vocalization (i.e., speech) riffs on an enduring benchmark for the detection of "intelligence" in our broader cultural imagination. The development of human infants is often judged according to when they vocalize their "first word"—an exciting moment for the family, especially if uttered between the age of twelve to eighteen months (Children's Health, n.d.), which is then interpreted as the infant being on a healthy trajectory of cognitive development. Turing's (1950) notorious "imitation game," now more commonly referred to as the "Turing test," singled out a machine's capacity to emulate human conversation as an adequate replacement for a machine's capacity "to think." Even in fictional media, a defining quality of so-called AI characters such as Samantha in *Her* (2013), TARS in *Interstellar* (Nolan, 2014), and HAL 9000 in *2001: A Space Odyssey* (Kubrick, 1968) is that despite their widely varying mechanical "bodies"—some remain in software (*Her*; Jonze, 2013), others are housed in metallic shells that can physically navigate the world (*Interstellar*; Nolan, 2014), and some are even given humanoid features such as expressive faces and limbs (*Ex Machina*; Garland, 2014)—they *always* have humanlike voices and linguistic capabilities. Indeed, the broader imaginaries through which we make sense of and envision artificial intelligence have been intimately tied to the symbol of *speaking machines*.

In this chapter, I shift my attention to these so-called speaking machines, or what I refer to as "AI voice engines." An engine generally denotes a machine that transforms one form (or forms) of energy into mechanical energy. Here, I deploy the term in the broadest sense to describe a system that requires and ingests human labor to automatically produce audio that sounds like a human voice. As such, these AI voice engines include systems that resemble not only the "now-hearing-and-speaking-ChatGPT," such as AI voice synthesis and voice cloning technologies, but also voice editing AI tools, such as accent modification[2] or spoken-language translation applications.

Synthetic voices generated through these AI voice engines are becoming part and parcel of our broader social and political infrastructures. They not only mediate everyday call center interactions and smartphone commands but are also used to enable key political and cultural exchanges. In the realm of politics, for instance, New York City Mayor Eric Adams "spoke" to his constituents in Mandarin, Urdu, and Yiddish using AI language translation tools (Coltin, 2023). In popular media, podcasters such as Lex Fridman interviewed world leaders such as the Ukrainian president Volodymyr Zelenskyy and India's prime minister Narendra Modi in their native languages. Criminal activities have also evolved: It was recently reported that members of the Trump administration and top business executives received calls from an individual who sounded like the White House Chief of Staff, Susie Wiles (Aratani, 2025). The list of headlines goes on.

These technologies, however, depend not only on speech-to-text (STT), TTS, and/or speech-to-speech (STS) machine learning models but also on the expertise, labor, and bodies of *voice actors/performers* who indirectly, and now, increasingly so, directly, contribute their recorded voices for the training of these systems. In this chapter, I turn my attention to these voice actors, who occupy a liminal position in the development of AI voice engines. As we will read, they must tread the line between "speaking machines" (actors speaking in mechanically legible ways) and "embodied humans" (humans adding valuable timbral variance to training data), as the embodied process of "voicing" is made commensurate with the statistical necessities of machine learning. If the previous two chapters documented my foray into the world of datasets, statistics, tables, ground truths, and machine listening as they relate to the *reading* of voices, this chapter brings the voice and all of its messiness back to the fore—to chart how voices are *engineered*.

Voice Actors

> With the rise of AI voice cloning, voice actors are being abused by the modding communities. As a member of the Skyrim modding scene, I am deeply concerned at the practice of using AI voice cloning to create and distribute non-consensual deepfake pornographic content.
>
> —@Robbie92_ [Twitter, now X, user], 2023

This viral tweet was a response to the emergence of pornographic content on Nexus Mod, a site that allows users to upload and download "mods"[3] for computer games. In this case, AI-generated voices trained on the voice recordings of real voice actors from the popular role-playing fantasy game *The Elder Scrolls V: Skyrim* were used to produce fan-made pornography. None of the affected voice actors had provided consent for their voices to be used in this way. But the availability of their recorded voices along with the accessibility of AI voice engines, such as VoiceLab by the company ElevenLabs, allowed modders to take and use the actors' voices to train AI voice engines to utter phrases in ways outside of the actors' control.

Professional voice actors/performers have historically been an overlooked and under-credited category of workers in the business of media production. They voice popular taglines, commercials, animated films, and video game characters that represent multibillion dollar industries but are rarely recognized outside of their specific communities (unless they also happen to be famous on-screen celebrities). The discrepancy between their compensation and their indispensability to these lucrative industries unfortunately also reflects this broader culture of exploitation. AI voice engines have exacerbated such discrepancies.

Tools such as VoiceLab can technically clone anyone's voice, but they work best with professional recordings done by skilled voice actors. This is simply because "clean" recordings (input) also produce "cleaner" outputs (Staniszewski, 2024). The expertise of voice actors is thus not only favored but also essential to the training phase and construction of ground truth datasets for professional AI voice generation models. Despite this, voice actors don't normally see any continued royalties or recognition beyond their one-off participation in the production of training data. It is not an accident that the voices of otherwise little-known voice actors from a popular video game were used for the production of synthetic AI voices. They presented valuable training data (free, clean, diverse, and consistent) to serve as the ground truth upon which machine learning models could "learn" what "voice" is and thus also replicate it. Indeed, the *Skyrim* modding controversy is not only a story of exploitation but also evidence of the indispensability of voice actors' expertise to the AI voice generation industry.

Critical studies of data have produced insight into the expansive human infrastructure of "data workers"—data annotators, content moderators, gig workers, ghost workers, crowd workers, platform workers, micro-workers, and so forth (Gillespie, 2018; Gray & Suri, 2019; Muldoon et al., 2024; Tubaro et al., 2020)—who make data possible and usable. In the context of AI voice engines, however, there is also the crucial step of *producing* voice data. Unlike text and image data, which are often collected by scraping the Internet, voice data for the purposes of developing professional AI voice engines need to adhere to more strict parameters of quality and consistency. Unlike "listening machines" that benefit from being able to "listen" in "noisy" environments, most voice engines don't benefit from producing "noisy" outputs. Most use cases for AI voice engines favor a particular type of output that resembles what people generally perceive as high-quality audio. As such, professional and directed recordings of voice actors are an important part of the data production/collection pipeline. Indeed, for machine learning practitioners, the voice is only useful insofar as it engenders materials sites from which sonic data can be mined. But voices (and the sonic data they generate) can't exist without the bodies that produce them. To understand the broader workings of AI voice engines, it is thus imperative to look beyond the machine learning practices of computer scientists and engineers and attend closely to the performers who lend their voices to this process.

Interviews

I spoke with ten voice actors for one to two hours each from a wide range of backgrounds and with varying years of experience to learn on a fundamental level their relationship with the voice. I found all the voice actors through a website and platform called Backstage that connects talent spanning artists, models, actors/performers, production crew, and post-production staff. As a more specialized platform that requires background checks from all users and is optimized toward media production, Backstage provided a safer and more reliable venue through which to get in touch with working voice actors and to tailor my search. I intentionally reached out to actors spanning a wide range of self-reported races/ethnicities, ages, genders, years of experience, self-identified voice "types," and performable characters and accents.

I also drew from my own twelve-year experience as a producer and songwriter, in which I worked closely with numerous vocalists in recording, directing, and producing vocal tracks, as well as my experience as a cultural adviser/consultant for a major American video game company, in which I provided input on voice actors auditioning for a South Korean game character, to guide my questions around vocal acting/performance. Throughout our conversations, we discussed how voice actors manipulate their bodies to produce certain effects/affects; how they leverage their own memories and experiences to channel certain characters and emotions; how they understand, maintain, and perform "authenticity" in the context of vocal performance; how they work with recording setups to produce "clean" recordings; and their overall philosophy to voice acting.

In addition to engaging with these voice actors as experts (in the case of experienced actors) or students (actors earlier in their careers) who could teach me about their craft and relationship to vocal production, I also approached my conversations with them as an important and rare opportunity to re-center their authority in this artistic-technical practice. Indeed, "voicing" is more than the mere piecing together and sounding of phonemes. It requires bodies, memories, and imaginations, all of which together mediate the voices we ultimately hear. I thus give ample space to my conversations with voice actors to highlight the varied and complex approaches they navigate to produce the voices (and now voice data) we consume.

In addition to my conversations with voice actors, I also spoke with a cofounder and several employees across two of the leading companies in the field of AI voice generation, ElevenLabs and PseudoVoice AI.[4] One of the employees I spoke to was a voice actor herself and transitioned into her role at PseudoVoice AI in 2021. She occupied a critical intermediary position between voice actors and the data infrastructures underlying AI voice engines as she was directly in charge of recruiting and directing voice talent to record voice samples, as well as cleaning that voice data to train the company's STT and TTS models. Through my conversations with other employees at various levels of development and management, I learned about not only the training and development processes of their AI voice engines but also what these startups viewed as the potential for AI voice generation and their business aspirations moving forward.

The Natural Voice

Across the voice actors and tech workers I interviewed, the voice existed as multiple things: as self, as expression, as tool, as performance, as data, as therapy, as resource, as opportunity, and as capital. But a consistent theme throughout these conversations was the importance of a "natural"-sounding voice. Only the ability to voice a character "naturally" would land a gig for a voice actor. Training data had to be cleaned and parsed to ensure vocal samples sounded "natural" for the TTS and STT models. Having a "naturally" deep voice rendered that voice actor more eligible for a particular gig than a voice actor who did not. Indeed, for voice to be meaningful and to serve the purpose of each of these different frames, sounding "natural"—whether innately, through refined performance, or with the help of technological editing—appeared to be an indispensable benchmark.

The core question, of course, is natural to whom? Naturalness (for lack of a better phrase) is a relational construct: To label something as natural is to situate that "thing" in a particular context with values, beliefs, and subjectivities. In this way, a natural-*sounding* voice is different from simply a *natural* voice. The first centers the listener, whereas the second centers the speaker. And what sounds natural to the listener is not necessarily what feels natural to the speaker. To produce a natural-sounding voice, then, is always to imagine and cater to a listener and the world that the listener inhabits. In the words of Eidsheim (2018), we must "listen to listening" (p. 27).

In her work with voice teachers, Eidsheim (2018) shows how vocal timbre is natural*ized* for classical vocal artists through decades of daily formal and informal vocal training that physically reshape parts of the body to produce certain timbres. She calls this process "entrainment" (p. 41). Here, voice teachers correlate natural timbre with vocal health, in which a "'healthy'-sounding voice is assumed to be a voice freed from blockages" (p. 45). These blockages can include any number of "bad vocal habits" (p. 45) as well as any other underlying physiological or mental conditions that are understood as "repressing" or "concealing" the singer's natural timbre. Indeed, a healthy voice is "assumed to be an unmediated sonorous conduit for the subject" (p. 46) that "correctly" bears the sonic markers of a singer's "natural" timbre.[5] The work of the voice teacher thus becomes a process of "uncovering" (constructing) this naturalness. Indeed, the natural-sounding voice is a product refined through years of consistent entrainment on behalf of the singer informed by a voice teacher's subjective mode of listening. The natural-sounding voice is *made*. And it is made according to how the instructor sees and listens.

Naturalness, in this way, is tied intimately to a *belief* in voices being consistent with the bodies (or imagined bodies) from which they are voiced. As such, entrainment can be understood as a process in which a vocalist restructures his/her/their body to produce a voice that is consistent with the body perceived by a listener. Here, there is a collapse of the social and the physiological, in which the vocalizer's vocal anatomy is literally reconfigured through training sessions—not unlike the way one builds muscle to change one's structural physique in a weight-lifting gym—in such a way that the produced voice "correctly" maps on to what the discerning listener understands as natural. The body is used to produce the voice, but the voice is also used to validate the body.

I similarly recount the different techniques used to produce natural-sounding voices by voice actors and tech industry workers. Voice actors have a special relation to the body, in which they not only imagine the voice as sounding different *parts* of their bodies (throat, head, stomach, etc.), but also manipulate or entrain their bodies in "unnatural" ways such as curling up into a ball or overly puffing one's chest out to physically emulate the types of bodies that they imagine certain voices emanate from. The ways they manipulate their bodies, however, are also influenced by their own memories, the imagined and/or literal audience, narrative interpretations, and personal experiences through which they create temporary worlds and personas for the purpose of embodying naturalness to themselves, their employers, and their audience. These are all further mediated through the physical and practical constraints of the recording setup. In this way, these processes of "making naturalness" shed light on how a natural-sounding voice for voice actors is always a practical, embodied, and tailored product born out of a purposeful collaboration between vocalizer and listener.

On the other hand, the tech industry workers I spoke with primarily saw themselves as managers or technicians of text and audio data. Primarily working with recorded voice samples, they described their work as refining these recordings so that they would appear "natural" and "clean" to the STT speech recognition machine learning models. These STT models could then produce clean text transcriptions, which would then be used to generate natural-sounding voices via their TTS voice generation models. Here, a natural-sounding voice is that which is both *legible* as a STT input as well as *producible* as a TTS output. Indeed, in contrast to the situated and embodied experiences of voice actors, the machine voices produced through TTS models are by design distributed, multiple, and abstract. The objective is to create composite "average" voices that depend on the bodies of specific individuals but ultimately transcend them.

In the sections that follow, I chart how natural-sounding voices are defined and made across these different communities. Whereas voice actors understand and construct the natural voice through the specific contexts in which they produce it, the tech workers I interviewed define naturalness as an engineerable technical feature that expands the market value of "voice" as a media commodity. Voices have been considered important media commodities for as long as sound reproduction technologies and mass audio communication infrastructures have existed. But synthetic AI voices transform the nature of their commodification. Indeed, synthetic AI voices become flexible "types" to be bought and adapted for various use cases. Naturalness is no longer a property of individual authenticity, expertise, or craft but rather a mechanical average that emerges through the curated processing of *multiple* voices stemming from *multiple* bodies. What is ultimately produced is a disembodied, distributed, and mechanical naturalness that both replaces the individual voices it is derived from and transforms fundamental understandings of what voice is and how it is to be valued.

Real and Made Memories

> The voice is only one part of a whole. You can't create the voice without creating the whole.
>
> —Tamara Becker Cimmerian, interview, 2023

Tamara started in professional theater as part of a traveling troupe since the age of seven, and professionally started using her voice as a singer in high school. Over the past twenty-five years, she has been working with her voice for records and film, and the last ten years have specifically been focused on voice acting. In short, she is a master of vocal performance. Throughout my hour-long interview with her, we talked about her approach to developing characters and their voices. Here, she showed me rich impromptu performances of how she switches into different personas and imagined settings to utter phrases in voices that sounded drastically different from the "out-of-character" voice she used to speak with me. I asked her about her process of constructing these different voices: Where is the source of that voice? How do you decide what sounds natural for a particular character or scene? Where are you pulling these emotions and vocal characteristics from? For Tamara, this was a process of world-building and storytelling. She explained that in order for her to be able to voice a character, she first needs to write the character's story. What was the character's childhood like? How was her relationship with her parents growing up? What was she lacking, and what are her basic needs? Does she have a need to be seen? A need for control? A fear of love? What drives this character? She emphasized that an individual's voice is a reflection of these experiences and memories—a voice without a story, according to Tamara, is not an authentic voice.

Tamara contrasted this method of world-building and "becoming" with another method called "substitution," which is a technique most commonly associated with method acting. She described substitution as drawing from one's own memories of sadness—for instance, watching a sad movie—and substituting that memory and the feelings associated with it to project a sad voice.[6] Tamara expressed that this method is not as authentic as "becoming"[7] or living the character because in order for a voice actor to truly vocalize the feelings associated with, for instance, "losing a daughter" (this was an example she gave based on a role she had previously voiced), she needs to truly believe that her daughter is dying. This voice is not simply a "sad voice" but a much more complex and intense one that is specific to "the uniquely painful experience of losing a daughter." For Tamara, if she were to substitute this intense feeling with a different "sad" one, it would produce an inauthentic voice.[8] Although personal experience and trauma are important channels and resources to be able to understand emotions fundamentally, she emphasized that these were only references from which she draws upon to *build* the memories and psychology of the character and not *substitute* them. In this way, she described the process of voicing a character as an extended process of learning about and becoming the character. For Tamara, a natural voice is ultimately the reflection of an authentic story, imagined or real, that she creates. Internal storytelling is part of the artistic-technical process of voice acting.

The relationship between naturalness and authenticity, and anchoring that relationship in a vocalizer's particular values, drives, and needs was a point that another one of the less experienced voice actors I interviewed also emphasized. Nick Banks had only been in the voice acting industry for six months at the time of our conversation in August 2023, and he still considered himself a student of the craft. Having worked previously in the nonprofit sector as a program manager and counselor, however, he had extensive experience in public speaking and mentioned that he frequently got compliments about his "naturally" deep voice. He thus took on freelance voice acting as a side gig, and learned quickly that his deep voice was one that was highly coveted for military advertisements.

Nick told me he never even considered these jobs because they didn't feel *authentic to him*, even though he realized that his voice might *sound authentic to the content or messaging*. Of course, the idea that a deep voice is somehow authentic to military ads is not necessarily a reflection of any natural connection between a lower vocal register and the military; rather, it points to broader social constructions of and relations around masculinity, the military, and their apparent biomarkers (of which a deep voice might certainly be one). But the point is that, for Nick, authenticity and naturalness were not rooted in the imagined listener, who might indeed make those associations, but rather in the vocalizer: himself. Describing the voice as "an internal rhythm" onto which the "soul is projected," Nick expressed that, for him, the voice is an expression of truth—not factual truth, but an internal truth. He thus mentioned that he finds it difficult to voice characters (as opposed to doing voice-overs) because it often requires him to step out of his lane, which can feel inauthentic to him. He told me that this is why he prefers to do voice-overs for commercials or content in which he feels like he is speaking in his "own natural voice." This might be attributed to the fact that Nick is a developing voice actor still in the process of learning how to use his voice, as well as understanding his relationship with it, but it also presents a conceptual entry point to interrogate the source of authenticity and naturalness in voice. How does authenticity map onto the multiple?

Jane Xi[9] is an author, therapist, and dancer who had officially ventured into professional voice acting two years ago at the time of our interview in August 2023. Jane told me she grew up in poverty and had to be her own friend for large parts of her childhood. She thus "learned to make things fun" by doing multiple voices for herself and sometimes her family such as the T-Rex from *Jurassic Park* or Gollum from *The Lord of the Rings*. Jane said voice acting became a coping mechanism for self-exploration and loneliness throughout her childhood and into her adult life. Discussing her struggles as an Asian American growing up in Boston, she said that she wasn't "given a choice to define [her] own individual perspectives or personas because of how Asian Americans are portrayed in media." She felt stereotyped into certain roles and personalities and was often told to police her own voice when she "cussed like a sailor" or spoke in a more aggressive Boston accent, which to her *felt* natural but to her community *sounded* jarring.[10]

For Jane, voice acting thus provided a way to explore parts of her identity that she felt were repressed by/in American society. By defining and articulating a voice within the parameters of different characters, "you can [also] bring that part of yourself out, which you may or may not have been aware of prior to that." She emphasized that growing up in a culture "where it wasn't cool to be Asian" precluded the opportunity to explore herself beyond what was projected onto her, and voice acting was a way to liberate herself from that past. She gave the example of voicing an overtly sexual ad for a radio station: "Maybe people didn't see that in you; maybe you didn't see it in yourself, but through that process, you dig into areas of yourself to bring those different aspects out." Indeed, as somebody who felt both invisible—as an Asian American—and hyper-visible—in moments where her voice didn't match her Asian body—Jane expressed that there is something liberating about voicing an unseen character, where one can escape the shackles of race, culture, ethnicity, and even species and the numerous stereotypes and beliefs that accompany those identity categories.

Like Tamara and Nick, Jane also emphasized the importance of being authentic to oneself in voice acting. But for her, the natural voice was also something that could be found through exploration. Unlike for Nick, who liked to stay close to what he understood as his "own natural voice," Jane's natural voice thrived in the multiple. She echoed a sentiment that Tamara also expressed, which was that "voice is truth, and voice changes all of the time because truth changes all the time; you grow, and your voice grows with you." Jane thus resisted the notion of "vocal identity," saying that a vocal identity feels sociopolitical and assumes a stable identity or label, whereas "voice" is more intimate and fluid—"it's a reflection of who you are in the moment, and that can change and does change all the time." For Jane, the voice served a therapeutic purpose, and it was within and through that process of self-exploration to which the definition of naturalness adapted.

A key theme throughout my conversations with Tamara, Nick, and Jane was the centrality of the self in the process of vocalization and one's ability to be in tune with one's own imagination, morals, values, memories, and experiences. Regardless of whether they were voicing characters or doing voice-overs, a vital part of the process of vocalization was understanding the context and then aligning themselves psychologically, morally, or from a position of therapeutic self-discovery. For these voice actors, the natural voice was first and foremost a personal feeling and a resonance of a larger and greater *whole* that exists, needs to be uncovered, or is imagined within themselves. In this way, they prioritized the vocalizer's subjective experience and process of voic*ing* as opposed to the sonic output of the actual voice. For Nick and Jane, this was what differentiated human voices from AI-generated voices,[11] both of whom expressed that to think of voice outside of this deeply subjective process misses a fundamental component of what voice truly is. According to Jane, the magic of intimacy, warmth, and authenticity in voice happens in the grounded and embodied moments of serendipity and improvisation, and that is why "humans will always have the upper hand."

Tamara, Nick, and Jane emphasized the centrality of *feeling* natural in the process of voicing and, to a certain extent, implied that this natural feeling would inevitably also lead to a natural-sounding voice. They also acknowledged, however, that this personal feeling *alone* is sometimes inadequate for producing a natural-sounding voice in professional settings. Indeed, as indispensable as that feeling of naturalness is to the production of voice, it is also arguably one that almost all speaking individuals feel in their everyday vocal interactions. This does not, however, necessarily qualify those individuals to be professional voice actors.

Mo Mellady, who has been a voice actor for more than thirty years and was also the voice of Erin Esurance, the pink-haired spy from the advertisement cartoons for the insurance company Esurance between the years 2004 and 2010, emphasized to me that voice acting is as much about feeling comfortable in your own voice as it is about understanding who you're talking to: the listener. Here, she described a process she learned from her own vocal education called MINE (Motivation, Intention, Need, and Emotion) that informed her process of analyzing scripts. This analysis revolves around the following questions: Who am I speaking to? Why am I speaking to this person? What is the need? What am I feeling as I speak to this person? What do I want them to feel as I speak to them? Mo mentioned that who she's speaking to completely changes how she decides to voice a particular line. And even if that audience is not explicitly defined like it is in the dialogue of a narrative-based piece—such as for a general voice-over, a radio tagline, or a commercial—it still requires her to conjure a figure to whom she is talking.

For instance, to return to Tamara's example of voicing a character whose daughter is dying, Mo pointed out the obvious fact that even after developing the world, psychology, and basic needs of that character, the voice can still vary drastically, depending on who that character is speaking to, and the naturalness of the voice will depend entirely on that context and relationship. In the everyday lives of speaking individuals, in which they are presumably vocalizing *to* somebody in the present moment, this is a tacit and taken-for-granted element of vocalization. Voice actors, however, are often vocalizing to an undefined group of listeners that traverses space and time. Mo thus emphasized that a key part of voice acting is to be able to actively imagine and construct the listener, even when there is none, at least in the moment of vocalization, so that her voice has a "direction" and serves the purpose of "connection."

Asking me if I ever get the sense that I can "hear a smile" over the phone—which she emphasized is more visceral than simply "sounding happy"—she explained that she can only "smile through her voice" if she can also imagine the listener as someone she likes and wants to "direct a smile to." According to Mo, she can't fake an audible smile as easily as she can fake a visual smile. Even if it's in the context of a quick voice-over, where she might not have made any kind of genuine connection with a listener, she said there's a deeper process of recalling intimate memories or thinking about loved ones that underlies her ability to truly speak in a way that feels and sounds as if she is smiling. This idea of voicing an audible smile captures the processes of voic*ing* that Tamara, Nick, Jane, and Mo all describe: a personal, embodied, intentional, and holistic action that is ultimately rooted in establishing a human connection between vocalizer and listener. As Tamara says, "The voice is only one part of a whole."

Embodied Techniques

> Trying to be natural always feels unnatural.
> —Christopher Wells, interview, 2023

Christopher was in the process of negotiating the terms of his first gig as a voice actor at the time of our interview in September 2023. He had been an on-screen actor for many years before he recently started to explore voice acting, and understandably, one of the primary skills that he had to move away from was his reliance on visual communication. He expressed that trying to convey the entirety of emotion and personality through the voice alone didn't initially feel natural to him. For instance, he became hyperaware of vocal habits and microphone techniques that would affect his ability to both feel and sound natural in his voice. He emphasized that a large part of his initial training as he transitioned into voice acting was thus learning how to repress many of his vocal and breathing habits by softening consonants and experimenting with air flow to construct different vocal timbres in ways that felt exaggerated or different from the way he was accustomed to voicing lines as an on-screen actor. He also mentioned that he learned to be hypercognizant of his proximity to the microphone. As an on-screen actor, he had to adjust the volume of his voice, depending on how close the camera was, but the much more sensitive condenser microphone he was speaking into as a voice actor meant that even little changes in timbre, breathing, and volume were amplified.

For Christopher, this presented a situation in which he was constantly dancing between *feeling* authentic—which would entail not being too conscious of these minute details—and *sounding* natural—which would require him to be completely aware of both his own vocal habits and the physical recording setup. He pointed out that some of the vocal habits he had to "correct" were part of what he considered his unique "natural" voice. But he was directed by the engineers he worked with that these natural elements of his voice did not necessarily translate to a natural-*sounding* (read: pleasant, consistent, clean, clear, etc.) voice from the perspective of the listener. In this way, constructing a "natural" voice as described by Christopher was a collaborative process with the audio engineer in which he reconfigured his bodily habits to produce a voice that better fit the conventions of a particular trade. Unlike Tamara or Mo, who emphasized the need to feel authentic *in order to* sound natural, the added element of adjusting to voice-only acting meant Christopher could only sound natural by somewhat embracing the *un*natural. As Sterne (2003) writes in relation to early sound recording, "Recording did not simply capture reality as it was; it aimed to capture reality suitable for reproduction. Spontaneity was spontaneous only through artifice" (p. 236).

A workaround for Christopher was to play with physical space. He mentioned that in contrast to a camera, a microphone allows him to do whatever he wants with his body as long as his voice produces the intended effect. By imagining the microphone itself as a person or sometimes even just an ear, he said that he can change his orientation toward and proximity with it based on these different mental frames. If he is experimenting with an intimate scene, for instance, he said he might imagine the microphone as the person's ear that he's whispering into, and he might get very close to the microphone in a way that would not be possible with a camera, which would then naturally open up his airways to allow for a softer and breathier voice. Here, like the voice actors from the previous section, Christopher relied on his *mind* to reimagine the space of the recording setup to prime his *body* in a way that would "naturally" produce the voice he wanted.

Tamara echoed these sentiments around microphone technique. In explaining her process of world-building and "becoming," she mentioned that she incorporates the unnatural but also intimate space of the recording booth itself as a channel for self-introspection. In her words, "It's not a space that you're ever in otherwise, but it can also actually help you be yourself." She described the process of blocking out the rest of the world through headphones, hearing her voice without any noticeable reverb, positioning her body toward the microphone in a way that reflects the situation she's voicing, and improvising in real time through the process of listening to her own voice through the headset as a required and unique component of voice acting. She explained, "[You're] weirdly talking to you, and the headphones literally make that happen because you hear your own voice in a way that you can never hear otherwise. You're literally not only the [talking] character but also the other person you're talking *to*" (emphasis added). For Tamara, the intimacy of the recording setup, which she described as a "little cocoon," allowed her to embody both vocalizer and listener, and this was an integral part of the voices that she produced as a voice actor.

In contrast, for Akinola Olabode, who had only been voice acting professionally for a few months at the time of our interview in September 2023, the recording booth represented a high-pressure setting. Akinola elaborated on the anxiety-inducing situation of being alone in the booth with people watching him from outside of the glass, hearing every minute detail of his voice through the headset, and understanding that he only has a few attempts to get it right. He emphasized that this often makes it difficult to get into character quickly, and the anxiety from that situation resonates in his voice. Listing simple techniques such as drinking water to moisturize his drying throat and lips, taking deep breaths, and closing his eyes, Akinola emphasized practical ways to physically prepare his body for vocalization.

Indeed, Mo also emphasized that mastery of the body is a fundamental requirement for mastering the voice. Mo describes our voices as products or resonances of our unique bodies, which she admitted is not necessarily how we're used to thinking about the voice in our everyday lives. "We're comfortable in our own bodies, so we're also comfortable in our own voices." Voicing a character, however, might require manipulating the body in ways that might feel unnatural. This feeling of unnaturalness is "natural," according to Mo, because voicing a character requires one to not only believe and imagine the *psychological* world of the character in the way that Tamara describes but also *physically* become the character. Here, she emphasized the importance of "gesticulation," the dramatic use of physical gestures to accompany the voice. In producing the voice of somebody who feels disgusted, for instance, Mo physically pretended to pick up a dead rat from the ground, which changed her facial expressions, contorted her body in a particular way, and even apparently sent chills up her spine. This was Mo's way of physically priming her body and mind to reflect those of someone who feels disgusted in order to produce that disgusted person's voice.

This is in stark contrast to how the dataset creators examined in chapter 3 conceived of affective or emotional voices, of which "voice of disgust" was certainly one. There, the premise was that there was a distinct acoustic feature of the vocalized sound itself that bears the mark of disgust, which itself was also conceived of as a coherent and stable emotional category. Unlike Mo, for whom the voice was just one part of a particular collection of disgust-related feelings and reactions, which included facial expressions, bodily gestures, and physical sensations, the SER dataset creators divorced the sound of the voice from the body of the subject who *feels* disgusted. This makes sense from a logic of generalization: Only by abstracting the voice away from the particular subject and body can one establish a category that *generally* denotes a "voice of disgust." As I will later show, this notion of abstraction and generalization for the purpose of stabilization is crucial for the construction of voice engines as well. But it is fundamentally at odds with how Mo and other voice experts actually work with the voice. For them, the voice is first and foremost a localized, fleeting, subjective, embodied, and unique act of self-expression.

Jane had similar thoughts: "Different bodies go with different voices." In addition to echoing what many of the other voice actors mentioned, such as physically making herself bigger by puffing her chest when she does the voice of a larger character or curling up and getting closer to the ground to make herself physically small and force herself to look upward toward the mic when voicing a child or smaller character, she also talked about how the cultural background of the character could affect the physicality and gesticulation that accompanied her vocalizations. "When I'm doing an Asian dialect, I make myself more proper; I sit upright, and my diaphragm will be propped up. It's almost like a lesser version of when I voice a robot." How Jane adjusted her body and posture was thus influenced by both her own experience as an Asian American woman being siloed into particular stereotypes in America and her own subjective beliefs about how Asian people carried themselves. The apparent "Asian-ness" of her voice first started with a particular posture, which further stemmed from her personal experiences and cultural beliefs. Of course, the posture itself is not inherently Asian. It's Jane's personal understanding and experience of "Asian-ness" that informs the posture and then the voice.

Indeed, when Jane showed me her "robot voice," she similarly had a medley of bodily configurations that helped her produce it. Throughout our conversation, it also became clear that underlying those bodily configurations were more fundamental beliefs about what a robot *is*. For instance, when doing her "robot voice," she wore a blank stare, her shoulders became squarer, the rest of her body appeared to be more rigid, and her mouth seemed to move only to utter the words, devoid of any expressive features. The cadence of her speech also changed noticeably. It was admittedly what one would expect to see and hear in a stereotypical "robot voice," but it was impressive nonetheless. She explained that she was operating from the perspective that a robot would be "human-like, but still unhuman." This affected, for instance, how she understood punctuation and meter:

> A speaking computer will not have variances in tempo nor will it have a recognition of punctuation to the accuracy that a human reader would. It won't necessarily be monotone, but it would also talk in a steady rhythm that doesn't represent the way humans talk. The point is to sound almost like a human, but not quite. Something needs to be a little off.

This belief of the "almost-human" robot thus informed her interpretation and execution of body and speech. She needed to first have her own beliefs about what a robot is in order for her to create the body from which a "robot voice" would be projected.

These subjective beliefs around body, voice, and identity comprised a central theme across many of the voice actors. Christopher, for instance, also emphasized what he called the "geometry" of the character: Which part of their body is resonating? What is the shape of their mouth and lips, especially for cartoon characters? Do they have a square body, a football-shaped body, or a pointy body? Recounting his experience working with a colleague who was voicing a zombie, he said that this colleague emphasized physically hanging the mouth and tilting the head to one side, having little control over her arms and legs, and letting her eyes droop as essential physical components to her giving voice to the zombie character. These characteristics, of course, are largely mediated through already existing media representations of how a zombie moves, similar to how Jane's beliefs about humanoid robots noticeably resembled cultural representations of robots in literature and film.

In this way, these processes of embodiment and imagination shed light on the interwoven relations between voice, body, mind, and culture. The voice is a resounding of these broader interconnected beliefs about the different sources—physical, psychological, cultural—from which the voice is produced. To emphasize the body in the production of voice is thus not to reduce voice to vocal anatomies but rather to understand voice as a unique instance of present and embodied self-expression, as all of the voice actors I spoke to described. Taken together with how the voice actors in the previous section described the process of memory recall and world-building, these conversations shed light on the intensely personal and situated nature of voicing from which memories and bodies cannot be divested. Indeed, the voices we consume as media products are contingent on these complex processes.

In the final section, I shift to my conversations with Mati Staniszewski and Tanja Milojevic, the tech workers who each work at two of the most well-known startups developing AI voice generation tools. I also bring attention to my discussion with Simogne Hudson, a (frustrated) researcher who used STT models on "unclean" audio. Here, I examine how the embodied processes of vocalization are folded into the production of synthetic AI voices as they are made commensurate with the statistical necessities of AI voice engines. Here, voices emerge as a fundamentally different type of media commodity to be mass produced. Indeed, as many of the voice actors I spoke to emphasized, to accept the outputs of AI voice engines *as voices* is also to rethink (reduce) what voice—and the act of producing it—is.

Synthetic Voices

> We want the real thing.
>
> —Mati Staniszewski, interview, 2023

ElevenLabs is a Polish tech startup founded in 2022 that specializes in developing AI voice engines, which include AI voice presets, AI dubbing, and AI voice cloning, among other voice-related applications. Despite its nascency, the company is already considered to be an industry leader[12] in the emerging field of AI voice generation largely due to the accessibility of its technology to general users, in terms of both availability and ease of use. Indeed, as mentioned earlier, one of its tools was what enabled the *Skyrim* controversy, in which existing recordings produced by voice actors for *The Elder Scrolls V: Skyrim* were used to train AI voices that sounded like the original voice actors for nonconsensual pornography.

Before companies such as ElevenLabs, AI voice generation was primarily a tool limited to technology companies that would offer voice cloning or synthetic voices as tailored products—that is, in collaboration with major studios for producing content such as *The Andy Warhol Diaries* (Rossi, 2022)—as opposed to an instrument available for general use. As most technology users are well aware, however, open/public access is always a double-edged sword. Indeed, although AI voice engines enable bad actors to take advantage of those tools in harmful ways, they also have the potential to allow users to consume content from across the world in their local languages. According to Mati, one of the cofounders of ElevenLabs who I had a chance to interview in August 2023, this was what inspired him to start the company. Growing up watching Polish-dubbed Hollywood movies, he said that it would have been a dream to be able to watch those same films with the "full emotions and plethora of voices" that exist in the original. For Mati, the Polish voice-overs didn't capture the "essence" of the original voices of the actors, and this severely took away from the artistic (and market) value of the films.

Imagine watching *Her* (2013) in South Korea and listening to Scarlett Johansson voice her lines in fluent Korean, in Tagalog in the Philippines, in Portuguese in Brazil, and perhaps even with a Kiwi accent in New Zealand. The premise is undoubtedly a compelling one.[13] And yet, it's also one that understands voice, speech, and language in a particularly modular frame (Kang, forthcoming).[14] What does voice need to be—and not be—to be transportable in this way?

For Mati, the voice must be abstractable, by which I mean the assumption that there is an essence to voice that can be abstracted from the original context of utterance. Only then can it be conceived that a voice that "sounds like Scarlett Johansen" (who to my knowledge is not fluent in Korean) voicing her lines in Korean is the "real thing." In fact, most multilingual individuals understand that voices and styles of vocalization change, depending on the spoken language (Bernhardsson, 2017).[15] This abstractive logic also constitutes a fundamentally different way of understanding voice compared with the highly situated ways that the voice actors I spoke to described. Indeed, to remediate voice through the logic of data *necessitates* reframing it as an abstractable, transportable, and stable object. As Raji et al. (2021) write in their systematic critique of benchmark datasets in machine learning, the production of an organizational ontology of real-world phenomena in the form of a dataset always results in a decontextualization and disembodiment of phenomena that are inherently linked to the subjectivities of those involved in the production of that dataset.

At PseudoVoice AI, this job falls on Tanja, the voice actor turned community/data manager I interviewed in September 2023. Tanja creates the ground truths for what voices are and how they should sound in the context of the company's AI voice engines. Tanja informed me that her job as a community/

data manager starts with recruiting talent who can provide the voice recordings that will comprise those ground truths. In the same way that I browsed through Backstage to identify a range of voice actors who had labeled themselves as proficient in particular styles of vocalization using labels such as "professional" or "young," Tanja also looks for voice actor profiles that fit the description for the type of AI voice she is collecting samples for.

During our conversation, for instance, she gave the example of a "sinister voice," which, according to Tanja, should have a "menacing, slow, [and] lower register." Once Tanja solicits voice actors to provide sample recordings based on the guidelines she provides them, she listens to the recordings and determines which segments align with the style that she is seeking, after which she deletes clips that "don't work." I asked her what her criteria were for determining what does and doesn't work. She responded candidly stating that she just "listens and subjectively decides." Her underlying presumption seemed to be that "you just naturally know if something sounds off."

Denton et al. (2021) contend with this logic of "naturally knowing" in their analysis of ImageNet, one of the most influential and widely used benchmark datasets for machine learning computer vision tasks. Here, they specifically critique the framing of the labeling process underlying ImageNet's construction as a process of "verification"—which "suggests the existence of an underlying truth about the presence or absence of a concept in the image" (p. 6). Indeed, when the actions of labelers are framed as "verification" as opposed to "description," this implies a universal mode of seeing that is detached from the specific subjectivities of seeing individuals. Here, labelers become interchangeable, in which "the human process of object recognition [is seen] as disembodied and decontextualized information processing, rather than a situated and contextual process of subjectively making sense of and describing the visual world" (p. 7). In this way, what underlies "ImageNet's epistemology is the assumed existence of an underlying and universal organization of the visual world into clearly demarcated concepts" (p. 6) that are also universally recognizable to all seeing humans.

For Tanja, the voice seemed to operate along a similar vein. A sinister voice was a "naturally recognizable" one, with particular sonic features described in terms of speed ("slower"), frequency/pitch ("lower"), and a self-evident feeling ("menacing"). This contrasts drastically with the way that voice actors such as Tamara described the process of voicing a character who lost her daughter, for whom the voice was only one part of a much broader context, psychological condition, and upbringing. For Tanja, however, the voice *necessitates* existing outside of the individual and only along abstractable and quantifiable dimensions. For instance, she emphasized repeatedly that one of her primary responsibilities is to work with her colleagues on the engineering team to make sure that the voice samples she gives them "don't have any drastic pitch changes." She called these samples "garbage data." To create the most natural-sounding AI voice, PseudoVoice AI's model needs to be trained on hours of "clean" voice samples that are consistent. In Tanja's words, "There should be no serious pronunciation issues; speed and prosody need to be even; and the style should match the label; 'conversational' [for instance,] should sound conversational."

As most individuals who speak and hear are aware, however, human voices rarely stay consistent in those ways. Indeed, when Mo, Tamara, and Jane performed their different voices in our interviews, they would often switch between talking fast, talking slow, speaking in higher registers, and quickly moving to lower registers multiple times throughout their take. Although Tanja requires the expertise of voice actors who have the *ability* to control and articulate their voices in ways that are flexible enough to produce intended "effects" (e.g., sinister), she also needs them to stay within a consistent and relatively tightly bounded lane of vocal production that renders their voice recordings legible to the company's STT models. For the purpose of producing training data, vocal expressions are not in service of conveying a particular emotion or encapsulating a scene. Rather, they serve the purpose of adding controlled variance that otherwise conforms to a particular set of acoustic standards determined by the data collectors (designers). For the voice recordings Tanja collects to be useful in the construction of speaking machines, she also requires the performers she recruits to speak like machines.

In this way, the voice actors subsume the role of the interchangeable data labelers. Here, their voices are not unique, situated, embodied, and artistic expressions of themselves as described by the voice actors I interviewed. Instead, they must produce voices that are meant to be transposable, disembodied, and decontextualized representations/instances of a particular vocal *type* (e.g., "sinister voice") for developing the ground truth of a more generalized AI voice engine. AI voice generation can thus be understood as a form of vocal *typification*. Despite the rhetoric of possibility and opportunity that fuels the voice AI industry (and the AI industry writ large), AI voice generation is dependent on the replication of preexisting norms. If a "sinister voice" is only defined according to a particular listener's (or group of listeners') subjective understanding of voice, it precludes the possibility of creatively understanding vocalized sinister-*ness* differently. "Sinister voices" outside of Tanja's definition are impossible in a system in which the ground truth she curates is the *only* truth.

The development of speaking machines thus also always obliges a process of listening first. Here, I turn to my interview with Simogne Hudson, a carceral studies researcher and scholar who was at the time (November 2023) helping a nonprofit organization transcribe radio commentaries recorded in prisons using Otter.ai's STT technology. Unlike the AI voice engines developed at ElevenLabs or PseudoVoice AI, the system that Hudson used is an AI transcription tool that translates audio data containing spoken utterances into text. As such, it is closer to a "listening machine" than a "speaking" one. But it still illuminates the centrality of the listening human in coproducing the voices that are made legible to and produced through machine learning systems. This is especially salient when such systems make "mistakes" or read voices in ways that conflict with the humans developing and/or using the technology.

Hudson, for instance, mentioned that several of the technological "errors" and "problems" flagged by both Otter.ai's STT model as well as subsequently by her colleagues—most of whom were "older, White women"—were "basically just AAVE [African American vernacular English] or non-White patterns of speech." To Hudson, these moments of "error" represented crevices through which she could infer both the contours of Otter.ai's latent grammatical structure, as well as observe the ways in which human editors were urged to listen to and refine (or not) Otter.ai's transcriptions. In particular, she recounted an instance in which a recording that contained the phrase "I haven't caught no misconducts" was transcribed as "I haven't caught any misconducts." Hudson deduced that "the way he spoke wasn't legible to Otter because he was using double negatives which is a common pattern in AAVE, but might not register as grammatically correct to the STT model." Otter.ai presumably expected to find "any" instead of "no" in that phrase and therefore "heard" and transcribed it that way. She noted that these types of minor changes can get frequently overlooked, especially since the human editors who finalize the transcriptions are also often listening and editing through the filter of what one might consider "standard White American English."

Hudson did acknowledge, however, that she believes some of these small editorial decisions are often necessary—not to the extent of changing words, but perhaps through punctuation choices and minor grammatical fixes. To Hudson, this editorial work represented a delicate dance between staying as close to the original utterance as possible, while also understanding her work as a form of journalism and reporting, in which she wouldn't want to represent a person in an "undignified way." Hesitant that this might be read as a form of respectability politics, she clarified that if she was the copy editor for a major publication, she would never want to represent someone's speech or writing in an overly literal fashion. Because she knew that these transcriptions would, to a certain extent, serve as representations of the inmates on the nonprofit organization's website, she described her methodology of punctuating and editing the transcriptions as a form of compassion and care work. It was an opportunity for her to add heft to what the callers from prison were saying. In this way, Hudson's mode of listening, her beliefs around what constitutes textual legibility, and her knowledge of how these transcriptions would be used directed how she ultimately transcribed the callers' voices.

Hudson further emphasized that this was especially important because Otter.ai's STT model would only transcribe "about 30–40 percent" of the audio accurately. This meant she also needed to rely on her own knowledge as a carceral studies scholar to contextualize the content of the audio recordings and deduce prison-specific terms. Indeed, Hudson mentioned that "without a background researching prisons, a lot of the tasks would have been way too hard." Even for an expert like Hudson, there were instances in which she needed to conduct separate research around the recording because Otter.ai very clearly mistranscribed words. In these instances, additional research would allow her to deduce, for example, what the name of a prison or individual uttered in the recording—but indecipherable to Otter.ai—would be. There was thus a necessary level of gathering a priori information around the recording before she could engage in the actual work of editing the transcription. In this way, despite her attempt to stay as close to the recording as possible, Hudson also had to make the recordings legible, audible, and understandable to herself before completing a transcription. Her listening praxis, experience as a carceral studies researcher, and positionality as a Black individual were indispensable to the voices she transcribed.

Similar to how Hudson and the nonprofit workers corrected and produced transcriptions according to the vernacular and contextual structures they adhere to, Tanja vets voice actors and cleans voice recordings according to her own beliefs about how certain voices (e.g., a sinister voice) should sound. A natural-sounding voice at PseudoVoice AI is thus directly tied to Tanja's position as a listening subject who, like Hudson, has the authority to delete, edit, and represent the voices she curates. The underlying goal for Tanja, however, is different from that of Hudson and her colleagues. Her objective is not to amplify individual voices but rather to develop new composite ones. In the example she provided, Tanja must thus tread the line between finding voice recordings that adhere to her own auditory standard of "sinister-ness" with the conviction that her standard is a generalizable one, while also collecting separate (and different) instances of "sinister-ness" that can add variance to the training data. Indeed, it is a delicate process of *managing variability*: The voice recordings she ultimately routes to her engineers must be "different enough" such that they create a meaningful average but also "sufficiently similar" that they conform to a specific standard.

In this way, synthetic AI voices are by design *superficial*. This is meant not in a moral sense but rather in a literal one: AI voice engines produce "surfaces." They are not tied to specific *sources* from which "naturalness" or "authenticity" is derived but rather invested in the replication of *effects*. If vocal production for the human actor is a process that is sourced from an internal structure of memories and experiences as well as by adapting to the external recording environment and context, synthetic voice engineering is a process of compressing and reproducing past performances. As such, the development process requires collecting material instances of performances (audio data) through which patterns can be found and averages can be computed. If a "sinister voice" for a human actor is dependent on the actor's comprehension and analysis of a scene, a "sinister voice" produced through an AI voice engine is the composite average of many such actors' comprehensions, analyses, and resultant vocalizations. In lieu of considering context, the presumption is that enough examples of adequate "sinister-ness" will produce an adequately "sinister-sounding" voice (figure 4.1).

Ultimately, the objective is to transcend the stories and bodies from which voices are derived so that only the sonic data—that is, the computable information of the voice recordings—remain. The separate utterances dissipate into data objects as the interchangeable material for synthesizing vocal "types" removed from their original contexts of production. Reduced to an

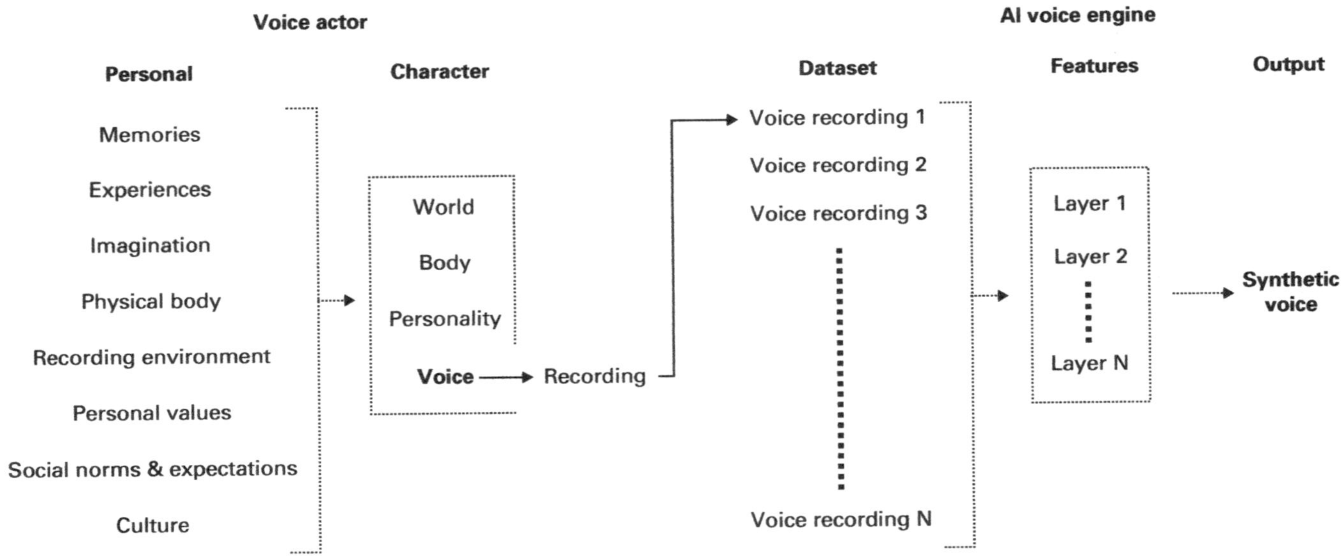

Figure 4.1
A simplified diagram comparing and connecting vocal production for human voice actors and synthetic voice generation for AI voice engines. It juxtaposes the embodied individuality of human vocal production with the inherently aggregate structure of synthetic voices.

effect or type, synthetic AI voices can satisfy the fantasies of scale, multiplicity, and efficiency that drive the voice AI industry.

* * *

AI voice engines reappraise the value of voices (and voice actors) as economic commodities to be mechanically mass produced. When voice is reframed and valued as a sonic effect, vocal production is also transformed into a simpler and more efficient process of audio production. Detached from the intricacies of human bodies and minds, vocal effects comprise suitable products to be manufactured, sold, and evaluated at scale.

The final chapter of this book delves deeper into the economics of synthetic voices and how they are managed (or not) in another world of numbers, charts, and gut feelings: venture capital. Indeed, a story of AI and technology development, writ large, cannot be told without the influence of risk-taking investors, grandiose promises, and techno-optimist futures (Andreessen, 2023a). As an area of artificial intelligence notoriously devoid of technical benchmarks, the domain of voice generation is one that is well positioned to not only think about the epistemic foundations of measuring "technical performance" but also reflect on the changing nature of machine learning evaluation in this unfolding era of generative AI. In chapter 5, I explore this topic of "AI evaluation" through technical benchmarking practices and venture capital investing.

5 Shifting Standards

Unicorns

Voice AI company ElevenLabs made headlines across the technology industry in early 2024 for successfully raising $80 million as part of its series B funding (Staniszewski, 2024). As reported by major tech journalism outlets such as Reuters (Hu & Cai, 2024), Forbes (Fink, 2024), Fortune (Confino, 2024), and Bloomberg (Bergen, 2024), this funding round, led by leading venture capitalists Andreessen Horowitz (a16z), Nat Friedman, and Daniel Gross and further joined by Sequoia Capital, Smash Capital, SV Angel, BroadLight Capital, and Credo Ventures, put the valuation of the startup at $1.1 billion, effectively granting it "unicorn status" in just two years of inception (Staniszewski, 2024).

The moment in which a technology startup reaches unicorn status is indeed monumental. And achieving it in such a short time frame is further reflective of the strong conviction by investors—predominantly venture capitalists at this stage—in the *future visions* as opposed to strong track record of the company. To provide some perspective, according to a report published in May 2023, the average time that it took current unicorns to achieve a $1 billion valuation was six years, with only ten companies ever accomplishing it within a year (Luchanka, 2023). Unlike more mature businesses that can showcase years of both quantitative and qualitative data documented in balance sheets, earnings reports, major personnel changes, journalistic coverage, and how the company has historically adapted to social, political, cultural, institutional, and technological pressures, a younger startup must justify its value and potential through *promises*, *prototypes*, and *projections* of potential market dominance—or what I refer to in this chapter as the "3Ps."

Because of this lack of historical data and the dependence on the inherently future oriented nature of the 3Ps, many VC-backed startups, even those that achieve the highly prestigious unicorn status, ultimately fail.

The most widely documented case of such a failure is perhaps "the blood unicorn Theranos" (Levine, 2018). Founded by the now incarcerated Stanford dropout Elizabeth Holmes—once labeled the "female Steve Jobs"—and the subject of several documentaries, television programs, podcasts, and a book, Theranos was a biotechnology startup that claimed to be able to "perform hundreds of tests (supposedly over 240) ranging from cholesterol levels to complex genetic analysis, with just a single pinprick of blood" (Roff, 2024). Whereas existing methods require a full vial of blood for each diagnostic test, rendering the process expensive, inconvenient, and inefficient, Theranos claimed to offer an easier and cheaper technological alternative. If successful, it would revolutionize the health-care industry.

Without repeating the now overtold story of the "decacorn's" (a startup with a $10 billion valuation) stunning fall, it should be noted that Theranos very much played the game that most emerging technology startups play to attract venture capital investments (minus the fraud). Coupling marketable charisma with grand promises, Holmes raised more than "$700m in investment from the likes of Larry Ellison and Tim Draper" (Roff, 2024), both of whom are notoriously shrewd and successful investors.[1] Indeed, the tale of the blood unicorn exemplifies, perhaps to an extreme extent, the consequences of when the 3Ps don't deliver—how grandiose promises end up being scientifically unviable, sleek prototypes crumble in practical applications, and enticing financial projections ignore crucial shortcomings of the business.

* * *

The final chapter of this book is about *progress*—scientific, economic, and the increasing imbrication of the two—in artificial intelligence research and development. Specifically, it attends to how progress is measured and thus defined in the context of the voice AI industry. Currently, benchmark datasets comprise the primary mode of model evaluation in machine learning. Broadly speaking, these refer to standardized datasets meant to represent a specific task domain (e.g., object recognition) that allow researchers and developers to test their machine learning models' performances on that specific task using a common set of reference points. As such, they provide standardized quantitative measures that allow different models to be easily compared (e.g., model A performs better than model B), as well as progress within a single model to be measured (e.g., model A is *x* percent better today than it was a year ago)—or at least that is the premise upon which they stand. Although such benchmark datasets exist for task domains related to computer vision and NLP, they are notoriously absent (or significantly underdeveloped) in the world of voice AI.

The absence of standardized benchmarks, of course, does not mean that voice AI systems forgo evaluation. Rather, different methods emerge to track progress, compare models, and justify investments. In fact, the recent explosion of "generative AI" models has further called into question the utility of machine learning benchmarking practices (e.g., Alaa et al., 2025; Eriksson et al., 2025), a contested custom that was already criticized for its narrow representation of tasks before the "generative boom" (e.g., Crawford & Paglen, 2019; Denton et al., 2021; Raji et al., 2021). In response to this changing environment, this chapter examines two modes of evaluation through which voice AI systems are currently assessed: (1) academic challenges accompanied by traditional benchmark datasets (less prominent in voice AI), and (2) venture capital appraisals and investment theses (more prominent in voice AI). As we will read, both are similarly limited in the worldviews they enact to measure and evaluate the "quality" of synthetic AI voice engines, but their consequences are vastly different. Given the growing dominance of industry research and development for artificial intelligence (compared with academic pursuits), it will be increasingly important to understand the different measures through which industry actors—namely, venture capitalists—measure, appraise, and influence the development of machine learning models beyond traditional benchmarking practices.

A Brief History of Venture Capital and Artificial Intelligence Research

Venture capital has always been a risky enterprise. In an expansive historical tracing of the development of VC in America, Tom Nicholas (2019) from the Harvard Business School finds striking parallels with the early days of the perilous whaling enterprise of New England. It similarly had to "grapple with the complexities of risk capital intermediation, organizational form, ownership structure, incentives, team building, and principal-agent tradeoffs" (p. 11). The resemblances are enough for Nicholas to single out the nineteenth-century American whaling business as the "origin" of VC-like investing practices in America. The most salient parallel would be the chasing of "long tails"—that is, relying on "hits"-based strategies:

> A detailed examination of the respective industries . . . shows how close the parallels are, especially regarding the distribution of pay-offs. The vast majority of VC-backed companies fail, making for a pronounced long tail. The "hits" are expected to offset the investments that yield losses and mediocre returns. The recent performance of a top-tier venture capital firm shows that 52 percent of the gross return on its portfolio was generated by startups that accounted for just 6 percent of the total cost of the investment portfolio. (pp. 12–13)

Many nineteenth-century whaling voyages similarly failed due to collisions, unfavorable climates, and even attacks. This often meant that more than 30 percent of voyages in a given year would generate a return of zero or below. This would be recuperated, however, by the top 2 percent of successful whaling voyages, which would generate profits in excess of 100 percent, a pattern that, as Nicholas shows (figure 5.1), uncannily maps on to the average performance patterns of VC investments (p. 14).

Despite the inherently risky nature of venture capital investing, the rise and continued dominance of the American tech sector, which has now become synonymous with the geographic region of Silicon Valley, cannot be explained without it. Indeed, "Silicon Valley [now] accounts for the vast majority of US venture capital funds, and most of the leading venture capital firms are located there" (Nicholas, 2019, p. 183). One of the conditions to which Nicholas attributes the establishment and continued success of the VC–tech industry partnership is the increased demand (and funding) for high-technology research and development by the US military in the mid-twentieth century: "Between 1955 and 1959, armed forces transistor acquisitions increased from $1.8 million to $99 million and the Department of Defense quickly became the largest consumer of such products. . . . Given that the military acted as an early adopter, it imposed strict technology performance standards on contracting firms" (p. 188). In other words, an important factor that allowed for successful VC–technology partnerships to emerge throughout the 1960s was both the high *technical bar* set by the US military and the guaranteed *market* that it provided for companies that met that standard.

There was one particular domain of science and technology research, however, to which this overall effect of military-imposed quality control

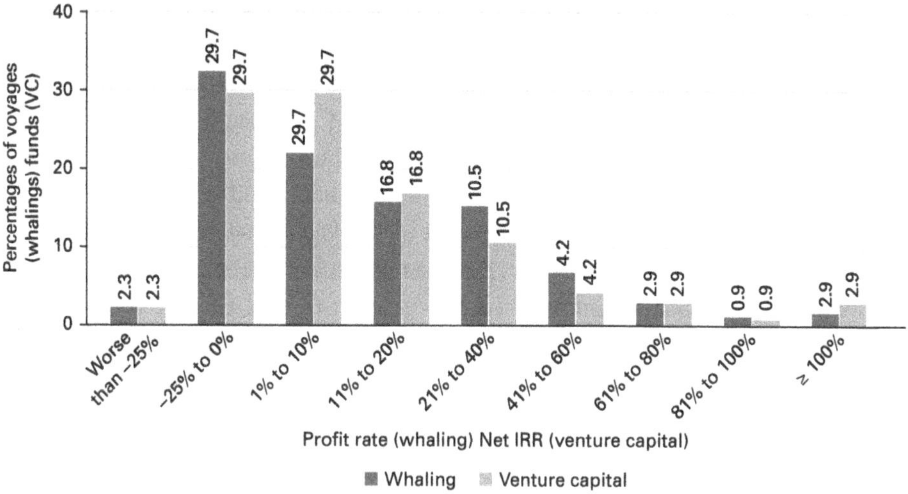

Figure 5.1

A redrawing of a graph adapted from *VC: An American History* (Nicholas, 2019, p. 14), which used data from *In Pursuit of Leviathan: Technology, Institutions, and Profits in American Whaling, 1896–1906* (Davis et al., 1997). It shows the profitability of whaling voyages plotted against the net internal rate of return (IRR) "realized by all funds in the VC industry across vintage years 1981 to 2006 tracked by Preqin, one of the major vendors of venture capital data" (Nicholas, 2019, p. 13).

and market making did not extend: artificial intelligence. To be clear, this is not to say that advancements in AI were not made during this time. In fact, the 1950s and early 1960s are widely considered the first ever AI summer. This was when John McCarthy first coined the term "artificial intelligence" at the infamous 1956 Dartmouth conference; Frank Rosenblatt built the Mark 1 Perceptron, an analog neutral network designed for image recognition in 1957; and Joseph Weizenbaum designed ELIZA, generally regarded as the first chatbot, in 1966 (Toosi et al., 2021). These technological innovations in early AI research attracted several millions of dollars in grants from the US military—namely, the Defense Advanced Research Projects Agency (DARPA)—throughout the early and mid-1960s.

As Will Orr and I write elsewhere, "Similar to the excitement around AI today, the hype surrounding AI research [during this time] created immense public expectations regarding the possibilities of the technologies; the hype [however] was similarly matched with exaggerations and false predictions by experts which further inflated expected capabilities" (2024, p. 1878). What differentiated the more exploratory academic research in artificial intelligence during this first AI summer from more hardware- and product-focused Silicon Valley technology companies was the absence of *testing*. Indeed, companies such as Magnavox, which "invented a public address system for battleships and anti-noise microphones for the US Navy's flying boats" (Nicholas, 2019, p. 187), or Ampex, "a company that would design antennas for military aircraft" (p. 188), developed technologies that could be tested directly on the field. Most artificial intelligence research at the time, however, was conducted by academic or academic-adjacent researchers at universities or labs as part of experimental research programs that focused quite literally on "cool demonstrations" as opposed to implementations of practical use.

Two critical reports published in the mid-1960s and 1970s—the *Language and Machines: Computers and Translations in Linguistics* report by the Automatic Language Processing Advisory Committee (ALPAC) of the US government in 1966 (National Research Council, 1966) and *Artificial Intelligence: A General Survey*, more commonly known as the "Lighthill Report," in 1973 from the UK (Lighthill, 1973)—also provided "grim prediction[s] for the technology's prospects" (Toosi et al., 2021 p. 455). In the absence of tests, evaluation metrics, and clear practical applications, funding for AI research could not be justified. This ultimately led to major cuts for AI research in the late 1960s onward. As a result, progress in AI development stagnated significantly, diverging from the continued success of Silicon Valley technology companies into the 1970s and 1980s.

Specifically, as Toosi et al. (2021) elaborate, there were three major issues with AI research during this time:

1. AI researchers were focused on developing systems that tried to replicate human thinking processes, writ large, as opposed to tailoring systems to particular tasks.
2. The complexity of real-world applications were oversimplified, which meant that most AI systems "succeeded mainly on toy (simplistic) problems, by combining simple steps to come up with a solution" (p. 455).
3. Expert opinions were largely unfavorable around the utility of neural networks and the limitations of their foundational architectures.

Simply put, the general sentiment after more than a decade of generous funding and excitement in AI was that the *promises* around humanlike intelligent machines, the early *prototypes* meant to embody those promises, and the *projections* for how these promises would revolutionize computing and society were largely underwhelming. The ensuing AI winter of the 1970s could thus be seen as "a crisis of trust and credibility" (Orr & Kang, 2024, p. 1879). Both the lack of explain-ability and transparency around progress in AI research in conjunction with its unclear applicability to real-world use cases meant funders could no longer justify support. The risks of the 3Ps were seen to vastly outweigh their potential rewards.

Standardization of Quantitative Benchmarks

A significant element that underlay the overall pessimistic sentiment around artificial intelligence at the time was the fact that standardized evaluation metrics for understanding progress in AI research were largely absent. Much of the funding during the 1950s and 1960s was provided to research labs "with few restrictions or accountability requirements such as the need for timelines or justifications" (Orr & Kang, 2024, p. 1879). This meant that vast amounts of resources were siphoned into AI research initiatives that more often than not yielded negligible returns.

In 1986, however, after more than a decade of mistrust from external funders, the tides shifted. Charles Wayne, a project manager at DARPA, along with Fred Jelinek from the International Business Machines (IBM) Corporation, introduced the Common Task Framework (CTF), which, among other details:

1. Provided a "publicly available training dataset with a list of feature measurements and a class label for that observation" (Raji et al., 2021, p. 3).
2. Provided detailed definitions of specific AI tasks.
3. Explicitly outlined quantitative metrics for evaluation for each task, in which the assessments would be "conducted by a 'neutral' entity, the National Institutes for Standards and Technology (NIST)" (Orr & Kang, 2024, p. 1879).

This framework served to address one of the primary concerns from funders at the time, which was the lack of any centralized and standardized method for evaluating progress. Considered "objective performance evaluations" (Liberman & Wayne, 2020, p. 23), "CTF measures became trusted proxies to represent a system's performance and general progression against standardized goals" (Orr & Kang, 2024, p. 1879). By making the largely closed-off and enigmatic world of AI research more open and accessible to nonexperts, the CTF effectively served to "create a basis for mutual accommodation in a context of suspicion and disagreement" (Porter, 1995, p. 149) between AI researchers and funders.

Although the CTF itself was primarily concerned with the development of such standards in the specific domains of speech recognition and machine translation, this standardized model of incrementally experimenting and evaluating against benchmark datasets quickly gained traction and started to cross over into other domains of machine learning such as facial recognition and gaming (Raji et al., 2021, p. 3). Eventually, the widely publicized Netflix Prize competition of the early 2000s adopted this framework, while also introducing, for the first time, the concept of a leaderboard for tracking progress on a massive scale across all of the participants.[2] This lucrative challenge, in which the number one ranking team on this leaderboard was to be awarded a million-dollar prize by the company, not only introduced industry actors as influential players in the development of artificial intelligence but also further cemented the authority of benchmark datasets, leaderboards, and accuracy scores/metrics in comparing, evaluating, and communicating technological progress in machine learning.

Indeed, this CTF model of benchmarking and standardized metrics-based evaluation has since become—for better or for worse—indispensable and central to the contemporary culture and practice of machine learning and artificial intelligence research, writ large. Kaggle, the data science competition platform owned by Google, hosts more than 630 ML competitions (at the time of writing in March 2024) that rely on this same logic of accuracy scores, benchmark datasets, and leaderboard competitions. Many of these competitions are funded by major research institutions such as the Harvard Medical School, the Bill & Melinda Gates Foundation, the Chan Zuckerberg Initiative, and Schmidt Futures, among others, and also offer large sums of money as rewards to the "winners." Widely attended annual "hackathons" or "challenges" in data science and machine learning such as the VoiceMOS Challenge (Huang et al., 2022), which I will elaborate on in the next section, also rely on this same framework.

Proprietary AI systems developed by major corporations and labs such as Google Deepmind and Microsoft's OpenAI also draw heavily from the CTF framework. Specifically, they use these benchmarks to demonstrate the capabilities of their models by reporting state-of-the-art performance on existing benchmarks. Consider this statement from the abstract of the paper published by Google DeepMind in conjunction with their multimodal AI model, Gemini:

> Evaluation on a broad range of benchmarks shows that our most-capable Gemini Ultra model advances state of the art in 30 of 32 of these benchmarks—notably being the first model to achieve human-expert performance on the well-studied exam benchmark MMLU, and improving the state of the art in every one of the 20 multimodal benchmarks we examined. (Gemini Team, 2023, p. 1)

Of the 136 words that comprise the abstract for the sixty-two-page document, fifty-nine (43.33 percent) are dedicated to communicating Gemini's SOTA performance on various benchmarks. A similar commitment can be observed in the abstract for the "GPT-4 Technical Report" published by OpenAI (2024), in which the authors emphasize that "GPT-4 exhibits human-level performance on various professional and academic benchmarks" (p. 1). Significant portions—arguably the majority—of the actual content of both papers are further dedicated to elaborating on the models' SOTA performances on these benchmarks, which the authors position as stand-ins for more "general" task domains such as "vision," "language," or "human-level understanding" (in representative subjects such as law or medicine).

The authority of benchmark performance in demonstrating model superiority—as demonstrated not only by both Google and OpenAI but also by the plethora of new models introduced at top machine learning conferences such as Neural Information Processing Systems (NeurIPS), the International Conference on Learning Representations (ICLR), and the International Conference on Machine Learning (ICML)—is well-documented. It should be made clear, however, that demonstrating SOTA performance on benchmarks does not necessarily translate to an AI model's (or a collection of AI models') general utility beyond its specific performance on a particular benchmark dataset.

Indeed, Raji et al. (2021) critique the examples of ImageNet (Deng et al., 2009) and GLUE (Wang et al., 2019) as particularly problematic in their broader acceptance by machine learning communities as "general domain" benchmarks—that is, proxies—for computer "vision" and "language," respectively. Their critique is based on the fact that these benchmarks, which are tied to specific evaluation metrics or tasks such as "object detection" or "sentence completion," are not necessarily representative proxies for expansive concepts such as "vision" or "language," which, of course, extend beyond such specific tasks. In the particular case of ImageNet, Denton et al. (2021) highlight contestable assumptions around (1) data annotation as an "objective" and "interchangeable" practice, (2) belief in a "universally true" taxonomy of the visual world, and (3) a near religious commitment to the scalar logic of big data as necessary preconditions for also accepting a benchmark such as ImageNet as a proxy for evaluating an AI system's capacity "to see." In other words, they argue that the efficacy of a benchmark in ascertaining an AI system's performance is only true given a particular set of preconditions, which makes their applicability to real-world use cases questionable. As Engdahl (2024) documents, these preconditions are also baked into the process of constructing benchmark datasets, thus further challenging their claims to wider task representation.

The centrality of benchmarks in demonstrations of AI model performance despite their questionable utility in representing the tasks they claim to measure, I argue, clarifies their role in the broader ecosystem of AI development: *communication*. Indeed, benchmarks are less technical measures than they are connective links or nodes that enable multiple channels of communication between different actors. They provide standardized grounds upon which quantitative comparisons can be made, which ultimately allow for the communication of superior technoscientific performance—of the technique, product, company, culture, and so forth. As such, there is a clear incentive to establish standardized benchmarks in any domain of artificial intelligence research as they provide a means of legitimizing that domain as a "measurable" and thus "progress-able" area of research and development. As both the original CTF model and the recent papers by Google Deepmind and Microsoft's OpenAI have demonstrated, communicating progress in this way is indispensable for attracting financial resources and promoting technologies.

In the sections that follow, I compare attempts by academic researchers in the voice AI research community to establish a CTF-style benchmark—in the likes of ImageNet (computer vision) and GLUE (NLP)—for synthetic voice generation AI models, with industry practices for appraising voice AI startups. In so doing, I analyze the shifting standards (from narrow technical performance to economic viability) through which generative AI systems are evaluated and developed.

Benchmarking Voice

> Human listening tests are the gold standard for evaluating synthesized speech.
> —VoiceMOS Challenge 2022

The above was the opening statement of the announcement accompanying Interspeech 2022's inaugural VoiceMOS Challenge, "a scientific event that aims to promote the study of automatic prediction of the mean opinion score (MOS) of synthetic speech" (Huang et al., 2022, p. 1). To provide some context, most speech synthesis systems, or what I refer to as "AI voice engines" in this book, are evaluated using listening tests in which listeners are presented with a collection of AI voice samples, after which they are asked to rate "each sample for *naturalness* on a discrete, 1-incremented scale from 1 (very bad) to 5 (very good)" (p. 2, emphasis added). As its name makes clear, the MOS is thus simply the mean (average) of the opinion (subjective) scores given by all of the recruited listeners/raters for a particular audio sample.

Traditionally, as Huang et al. (2022) write, these MOS tests were conducted on a situation-by-situation and system-by-system basis. This meant that the scores were not only incomparable between different systems (different listeners, different instructions for evaluation, different types of AI voice samples, etc.) but also extremely time-consuming and costly to generate. The inaugural VoiceMOS Challenge thus sought to address this obstacle by constructing a shared framework for speech synthesis researchers—it was a classic CTF-style response:

> In order for MOS prediction models to be a useful tool for speech synthesis researchers, these issues need to be addressed and more improvement is needed in the state of the art in the MOS prediction task. With these aims, we developed the VoiceMOS Challenge, a *shared task* using *common datasets* for MOS prediction. We conducted a large-scale MOS listening test to cover many different types of TTS and VC [voice conversion] systems spanning many years, and this dataset forms the basis of the challenge. (Huang et al., 2022, p. 1, emphasis added)

To circumvent the resource-intensive and context-dependent nature of existing MOS-type listening tests, this challenge is centered on the idea that it would be more efficient and effective to *predict* derivative MOS instead of actually *collecting* them. As the logic goes, if a model can be developed that reliably and accurately "learns" the patterns between the AI-generated voices included in the large-scale dataset compiled for this challenge and the associated MOS, not only would there be no need to conduct new listening tests (and derive new MOS) for every new iteration, but also it would establish a centralized benchmark against which all future AI voice engines could be tested. The goal is to develop an "objective"—read *not* as "accurate" but rather as "centralized," "standardized," "shared," and "distanced"—way to evaluate the so-called naturalness of AI voice engine outputs.

Before addressing the benchmark itself, however, it is important to linger on the challenge's derivation of an "objective listening metric." Indeed, the predicted MOS is a *derivative construct* extracted from past listening practices that is meant to stand in for a broader collective of human listeners "listening for naturalness." My use of the word "derivative" here is borrowed from the world of finance: "With derivatives, certain characteristics of a commodity, asset or financial security can be separated out, priced and traded in their own right. The wheat, the shares, the bond, the energy, and, of course, the weather themselves are not traded; just the attributes" (Bryan & Rafferty, 2006, p. 10). A prime example of a financial derivative is the *stock option*, which is a separately tradable "contract" that allows an investor to buy or sell a stock at a certain price in the future. The price of such a stock option is dependent on a particular attribute of the stock: the current and projected volatility of the stock price over the temporal span of the contract. As Bryan and Rafferty (2006) write, however, "Derivative prices dance around 'real' asset values" (p. 12), in that they are not *un*related to the price of the underlying asset but, in practice, can effectively circulate as a separately abstracted *thing*.[3]

The authors thus profess that the common understanding of financial derivatives, in which their values are seen as "deriving from the value of an underlying asset . . . [is] profoundly limited" (p. 11). Indeed, "in actual derivative markets it is apparent that prices do not in fact 'derive' from the original asset . . . On the contrary, prices usually run the other way, so that options and future markets are the places where prices are first formed, and prices in cash markets follow" (p. 12). In other words, the price of the actual stock can be—and frequently is—influenced by options traders' purchasing of stock options, which are disaggregated attributes turned into tradeable financial products meant to represent predictions of how much the price of the associated stock is projected to move in the future. There is thus a paradoxical reversal in derivation, in which a dynamic options market, built around a particularly abstracted attribute of securities, precedes and further influences changes in the actual stock market (and not the other way around). The derivative is not deriv*ed* but rather derive*s*.

In the specific case of the MOS listening test, a similar reversal of derivation occurs. It is no longer the embodied human practice of listening to synthesized speech, *nor* the individual scores of "naturalness" provided by the human listeners, *nor* even the average of those collected scores, *but rather* an AI model's *prediction* of the average of the scores meant to represent audible "naturalness" on a scale of 1 to 5 that becomes the derivative metric to evaluate synthesized speech. The structural inelegance of the previous sentence is intentional: It represents the similarly unusual and multiple levels of abstraction that must occur in service of the underlying premise of the VoiceMOS Challenge to derive an "objective" and mechanical form of evaluation from the inherently human and subjective process of listening. Not unlike how the stock option, *in theory*, is derived from the underlying security but, *in practice*, can influence the pricing of the security and actually drives the eventual movement of the associated stock price, the mechanically predicted MOS also, *in theory*, is derived from how a group of humans listen but, *in practice*, constructs an autonomous metric that paradoxically represents not a single embodied listener in its attempt to mechanically fabricate an aggregated simulacra of "collective objective listening."

If one takes a moment to step back, the situation—in the name of automation and efficiency, masked with the banner of "objectivity"—is comically techno-utopian:

1. Human voice actors produce recordings in ways that are legible to machine learning models along quantifiable parameters of "quality" (minimal noise), "frequency" (consistent tone), and "cadence" (consistent speed). Their goal is to sound as "natural" as possible, in which that naturalness is often produced through *un*natural methods and is defined through machine legibility.[4]
2. Data related to these parameters in the recording are disaggregated and reaggregated to synthesize natural-sounding AI voices.
3. Multiple human listeners evaluate the "naturalness" of these AI voices on a scale from 1 to 5. The averages of those subjective scores (MOS) are derived for each AI voice.

4. These separate listening evaluations are expensive and inefficient. Therefore, it is decided that even these evaluations should be automated. Instead of having humans listen and rate for naturalness, a machine learning model should "learn" how humans have rated different AI voices for naturalness in the past to predict scores for new AI voice engines.

The end result: The AI "ear" validates the AI "voice."

VoiceMOS Benchmark Dataset

At the center of this techno-utopian venture, of course, is the benchmark dataset constructed for the inaugural VoiceMOS Challenge (Huang et al., 2022). The authors don't give a name to this benchmark dataset, but they do mention that it is composed of samples from previous challenges such as the Blizzard Challenges (e.g., King & Karaiskos, 2010, 2011, 2013, 2016), Voice Conversion Challenges (Lorenzo-Trueba et al., 2018; Toda et al., 2016; Zhao et al., 2020), and from ESPnet-TTS (Hayashi et al., 2020). Following precisely the same design of the previously described MOS listening tests, Huang et al. (2022) asked listeners to rate the samples for "naturalness" on a scale from 1 to 5 (p. 2) to construct this challenge dataset (hereafter, the "VoiceMOS benchmark" for simplicity). In other words, their solution to their very appropriate critique of previous MOS listening tests—that is, that their utility is questionable because they are inextricably tied to the subjectivities of the listeners, the systems, and the contexts in which they are conducted—is paradoxically to *recreate* it, just *bigger* and tied to a *centralized challenge*.

This is strikingly reminiscent of the early days of computer vision research when the inaugural ImageNet Large Scale Visual Recognition Challenge (ILSVRC) was established in 2010. As Fei-Fei Li—one of the principal creators of the original ImageNet dataset—states in a 2017 talk titled "ImageNet: Where have we gone? Where are we going?," the goal of developing a "gold standard"[5] dataset for visual recognition was motivated by a steadfast commitment to, and particular assumptions around, human vision. Indeed, initial attempts involved hiring Princeton undergraduate students to provide text labels for images in the dataset based on the (problematic) assumption that uncontestable relationships exist between words and images—that is, that the way a particular group of Princeton undergraduate students "see" is how *everyone sees*. As Denton et al. (2021) document, the researchers did ultimately move away from using undergraduate students as annotators. But it was not because they realized the limitations of their assumption that all seeing individuals label images the same way. Rather, it was simply financially unfeasible to rely on the limited and expensive labor of Princeton undergraduates to complete the task. According to calculations by Jia Deng, one of the primary creators of ImageNet, "It would take 19 years to label the entire dataset that way" (p. 7).

The creators thus chose to circumvent this obstacle by leveraging the efficiency and low cost provided by the Amazon Mechanical Turk (AMT) crowdsourcing platform, which ultimately allowed contributions from "49 thousand workers from 167 countries" (Denton et al., 2021, p. 8). Through such tactics of scale (achieved via AMT) and centrality (via the annual ILSVRC competitions) along with the accompanying rhetoric to promote that scale and centrality in research papers, ImageNet effectively became the uncontested "gold standard" benchmark for computer vision tasks. This was despite the dataset's numerous problems around construct validity, bias, data quality, and representability, among others, repeatedly pointed out by peer researchers in the field (e.g., Birhane & Prabhu, 2021; Crawford & Paglen, 2019; Denton et al., 2021; Raji et al., 2021; Russakovsky et al., 2015).

In this way, the supremacy[6] of a benchmark dataset is achieved not necessarily through a thoughtfully curated design or structure but rather in its marriage of *scale* achieved through resource efficient methods such as AMT with *centrality/availability* often enabled by their use in field-defining challenges such as ILSVRC. The development of MOS-*prediction* models in the case of VoiceMOS can be seen as a step further (compared with the use of AMT for the construction of ImageNet) in the attempt to economize human-derived "gold standards." Here, those gold standards take the form of reductively derivative formats of "listening" enabled by displacing the act of listening from embodied human listeners to scalable ML prediction models. Listening is thus reconceptualized as a collective and mechanically "average-able" feature to be replicated via machine learning.

It is important to note, however, that despite the ambitious opening statement of the inaugural challenge as well as its undeniable similarities to the development of ImageNet and the CTF, the VoiceMOS benchmark is not *yet* a "gold standard" in the broader field of AI voice generation. Part of this, of course, is due to its nascency, which is why this marks a pivotal moment in the development of AI voice engines to observe how and if other academic researchers[7] and, perhaps more importantly, industry practitioners/developers are in conversation with these centralized challenges and benchmarks.

It is well known, for instance, that despite the lack of attribution, academically developed open-source datasets significantly influence the structure of how proprietary datasets are constructed (Orr & Crawford, 2024). Currently, it's not clear how privately developed AI voice engines such as those from ElevenLabs are evaluated and whether metrics such as MOS are engaged with or even considered in their development. In a conversation I had with Robin Netzorg, for instance, who at the time was a computer science PhD student at the University of California, Berkeley, conducting research on subjective evaluations of synthetic AI voices, she expressed that "MOS doesn't really tell you how much a synthetic voice is 'wrong.' It's more like a base check, where a low MOS score is definitely a red flag, but a high one isn't as informative." In other words, MOS is valuable as a baseline threshold and has historically been a relatively effective means of filtering out extremely "low-quality" voice generation outputs, but it might be less of a relevant metric in the context of ElevenLabs, in which tailored and perceptually more "realistic" synthetic voices are generated.

This does not mean, of course, that other metrics or standards don't exist. Similar to how machine learning benchmark datasets emerged in the 1980s as a way for AI researchers to attract and justify DARPA funding by showcasing quantitative progression on shared standards, startups such as ElevenLabs also need to communicate progress and model supremacy to their investors in order to successfully garner financial support. Here, it is not a centralized government but rather venture capital firms that AI developers must court. These VC firms are motivated by different incentives compared with government bodies, which makes it important to closely examine the AI industry's relationship with these companies and how its AI technologies are evaluated, appraised, and selected as candidates for investment.

Industry Evaluations

In February 2024, I spoke with Ghost,[8] an investor from the venture capital firm a16z, who was involved in the firm's series A and B investment rounds for the voice AI unicorn ElevenLabs. To provide a bit of context, a16z is one of the world's most powerful Silicon Valley venture capital funds with a notoriously bullish position on the future of artificial intelligence. One

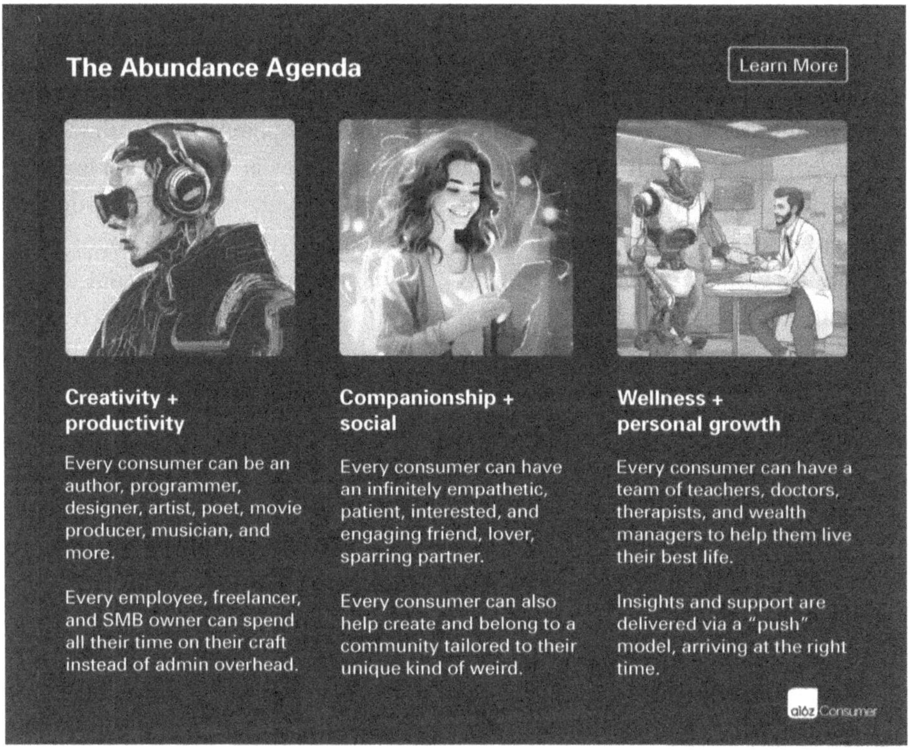

Figure 5.2
A screenshot from the a16z blog post "How AI Will Usher in an Era of Abundance" (Acharya, 2024) taken in April 2024. It describes the firm's techno-optimistic views around artificial intelligence for consumer products in what it calls the "Abundance Agenda."

of its cofounders, Marc Andreessen, famously wrote "The Techno-Optimist Manifesto" (2023a) as well as a blog post titled "Why AI Will Save the World" (2023b), both of which are largely centered around what Andreessen believes "AI *could be:* A way to make everything we care about better" (2023b). Separate from my own thoughts on a16z's broader outlooks on AI (most of which I find questionable, but a few of which I do agree with), the company's underlying commitment to what Andreessen calls "techno-optimism" is an important backdrop to understand the firm's investment theses in startups such as ElevenLabs (figure 5.2).

In the world of finance, an investment thesis generally "refers to a reasoned argument for a particular investment strategy" (Kenton, 2023).

Although there are no specific standards as to how an investment thesis is composed, most theses will primarily consist of cost–benefit analyses of the investment, potential risks and losses related to both internal company dynamics as well as macroeconomic trends, different potentials/avenues for growth, contents of financial statements, exit strategies, and analyses of leadership, among other factors. As an investment ideology, "techno-optimism" permeates a16z's approach to constructing their AI investment portfolio. In Ghost's words, "If a technology wins, its usually winner takes all, [and] the point is to bet on the one time that it works, [because] that covers the ten to twenty times it doesn't. Optimism is key. Optimism is harder and riskier than cynicism." In this way, Ghost's description of techno-optimism very much echoes the broader investor sentiments that Nicholas (2019) describes in his recounting of the risky and speculative New England whaling enterprise of the nineteenth century and the chasing of long tails. Indeed, even more so than traditional investing, venture capital is especially dependent on speculation because an adequate financial track record—a crucial indicator of a company's financial health and therefore growth potential—is missing.

According to Ghost, the "optimism" in "techno-optimism" largely denotes the possibilities around machine learning and scale: "Scaling laws apply—shove it [an ML model] compute and it has emergent qualities—[which means] generative AI is going to make the marginal costs of creation zero."[9] This apparent magic of scale that Ghost describes has been at the center of much of the renewed interest in artificial intelligence across the field at least since the "success" of ImageNet in 2010. More recently, however, leading AI researchers such as Google Deepmind CEO Demis Hassabis (Knight, 2024) and Meta's Chief AI Scientist Yann LeCun (Perrigo, 2024) have expressed skepticism around the so-called compounding returns of scaling *alone* in the context of compute and data. Yet, for Ghost and a16z, commitment to the scaling law at least within the context of domain-specific applications—that is, that domain-specific generative models will continue to "improve" in their capacities with more compute and data—is a driving factor in the firm's techno-optimistic position.

For instance, Ghost provided the example of developing a video game entirely with generative AI models. Here, Ghost expressed that the goal is to identify what "the primary building blocks of reality" are. Once those "building blocks" are identified (audio, video, text, etc.), the goal is to "go after modality after modality" because it's not guaranteed that "just because [a company] is good at image, [it'll] be good at sound." The thesis is rather simple: Because scaling laws ensure growth, at least within a specific modality, and each modality requires specialization, a16z is going to identify and invest in leading companies that specialize in each of the primary task domains. In Ghost's words, "No matter how you cut it, audio is a very important modality. One audio AI company will come out on top—no matter what." ElevenLabs is a16z's bet.

I asked Ghost how the company came to that decision. What were the indicators that pushed them to invest in ElevenLabs but not in other voice AI startups such as Resemble AI? It's clear that within a broader AI investment thesis driven by ideologies such as "techno-optimism" (Andreessen, 2023a and the "Abundance Agenda" (a16z, 2024), generative AI for audio is a central area to target. But why *specifically* ElevenLabs? What is the *company-* or *model-specific* investment thesis?

Traditionally, widely accepted machine learning benchmarks would serve this role in differentiating the performance of different ML models within a set domain. As we can recall from the documented history earlier, this was one of the primary reasons that machine learning benchmarks were developed in the first place: to communicate progress, differentiate between models, and therefore compete for funding during an era in which institutional trust and support for AI development was eroding. In this way, even though there has been countless critical research shedding light on the shortcomings of benchmarking practices in machine learning, it is undeniable that the benchmark itself has been and arguably still remains useful insofar as anchoring technological progress and comparison to a *standard* means of evaluation—irrespective of the *efficacy* of that method in representing the actual task.

For sake of comparison, one might understand this division between the practical utility of standardization and efficacy in evaluation alongside the context of standardized testing for college admissions in the United States. The SAT (previously known as the Scholastic Aptitude or Assessment Test) was dropped by many American universities as a requirement for college admissions primarily during the COVID-19 pandemic. This transition away from the SAT was also, however, somewhat due to increasing public skepticism around its utility in measuring an individual's prospects in college and beyond. Indeed, similar to the way in which "data overfitting" has been criticized as a particular limitation of using benchmark performance as a measure of machine learning models' efficacies on particular tasks (e.g., Orr & Kang, 2024; Raji et al., 2021), the proliferation of "a lucrative test-prep industry in the United States, which is estimated to grow by about 7% to almost $50 billion by 2027" (Rothenberg, 2024), also means that with adequate preparation, students can be optimized specifically for scoring high on the SAT. This confounds the broader premise of assessing a student's general/overall aptitude or scholastic ability.

Despite these widespread critiques of the SAT as a limited measure of student assessment, several of the most competitive institutions in the US such as Brown University, Dartmouth University, Harvard University, the Massachusetts Institute of Technology, Stanford University, and Yale University have all brought testing requirements back to their admissions process in recent years (e.g., Bregel, 2024; Carter, 2024; Lorin & Bloomberg, 2024). One of the primary reasons provided for this reinstatement is the lack of a *standardized* American education:

> Education experts who spoke to CNN say the SAT is necessary because there is no "standard" American education . . . Even within the same school and subject, some teachers might grade more or less leniently than others. College admissions officers simply don't have the ability to account for all of these minute discrepancies, experts told CNN, so standardized tests will likely stick around unless US education becomes standardized in other ways. (Rothenberg, 2024)

Indeed, similar to the justification for funding cuts during the first AI winter of the 1970s, the uncertainty and suspicion around the absence of a centralized and standardized mode of evaluation ultimately necessitates the development of external measures/standards to fill the vacuum. As imperfect as most standards are, they are arguably foundational for any kind of operation that spans across time and space, especially one that upholds notions of "fair evaluation" (Bowker & Star, 1999; Mulvin, 2021; Porter, 1995).

In the context of privatized AI product development and venture capital investing, however, especially regarding novel applications such as synthetic AI voice generation, widely accepted and readily standardized technical benchmarks are not always available. There are, of course, attempts to establish such standards such as VoiceMOS. But as Bowker and Star (1999) argue, a standard only becomes a standard through wide adoption, which VoiceMOS has not yet achieved. In such an absence, financial measures—that is, the investment thesis—can fill the vacuum and take on a significantly more central role in evaluating a startup and its AI products' growth potentials.

The investment thesis can thus also be understood as its own kind of industry/financial benchmark that is used to rationalize funding and therefore "enable progress" for model development. Indeed, according to Ghost, there were no technical benchmarks (such as VoiceMOS) against which ElevenLabs' (or other competitors') AI voice generation tools were tested. Rather, the evaluation processes largely centered on meeting with the leadership of potential AI audio startups and personally testing their products:

> We don't necessarily need a technical benchmark. We already know we want to own a generative AI audio company, so all we need to do is find it. An advantage of being at a mega fund like this means we can easily narrow down our search to a few of the most active companies and meet with the founders and team. We get access to all their products and then test them out personally. (Interview with Ghost)

The investing team itself thus becomes the momentary standard through which various technologies are evaluated, wherein the subject of evaluation shifts from the technical machine learning model (in the case of traditional benchmarking practices) to what I described at the start of this chapter as the 3Ps: promises, prototypes, and projections. As Ghost describes, their investment thesis was formulated based not on evaluations of various generative voice AI models against an established technical standard but rather on subjective judgments[10] of how committed founders seemed to their visions (promises), personal experiences experimenting with early-stage products (prototypes), and analyses of how present operations might grow into future gains (projections). In this way, the 3Ps reemerge as a principal means through which an investment thesis is formulated and therefore, as an extension, how a candidate technology is evaluated and ultimately takes shape.

Shifting Standards

There are key implications around this increasing shift toward a more explicitly subjective,[11] financially driven, and speculative form of evaluating technoscientific progress. On the surface, it can be seen as a reversion to the AI research practices of the late 1950s and early to mid-1960s, when "cool demonstrations" and early prototypes alone served as the basis for lofty promises and ambitious projections. A key difference, here, however, is the wide availability of *user metrics*.

In a post from June 2023 announcing their initial series A partnership with ElevenLabs, a16z investors wrote, "ElevenLabs has massed over 1 million registered users who have generated over 10 years worth of audio content" (Li et al., 2023). Unlike the 1950s and early 1960s, when early AI prototypes were not coupled with any type of quantitative metrics—thus necessitating the eventual development of the CTF in the 1980s—the prototypes of today, at least those developed by private companies, do fasten their promises and projections to market/user data. To a certain extent, these user metrics serve to replace the role of the quantitative performance metrics enabled by technical benchmarks. Ghost also echoed these sentiments: "Outside of personally testing the products and meeting with the team, we also look at revenue and web traffic. Both are good proxies for evaluating a company and its products. What gives us confidence is that developers are adopting it. Users need to adopt it too. This is more important."

Although the failure of the 3Ps during the early days of AI research spawned a competitive structure of standardized evaluation in the following decades through developments such as the CTF, the Netflix Prize, ImageNet, and the rise of leaderboards and competition benchmark datasets in venues such as Kaggle, we are now seeing a gradual transition away from "technical performance"' metrics toward a more capitalistic and market-driven type of evaluation couched in user adoption and profitability. This is especially the case in what might be considered more "subjective" task domains such as synthetic voice or image generation, which are inherently less compatible with standardized measures of assessment (Orr & Kang, 2024). Indeed, as we saw in the case of VoiceMOS, attempts to standardize the evaluation of such tasks result in awkwardly ambitious abstractions (distortions) to fit the task to the evaluative framework (and not the other way around).

Even in the case of GPT-4 and Gemini, which, as I recounted earlier, were released alongside lengthy scientific papers that detailed their SOTA performance on benchmarks spanning various tasks, their capabilities and success for the general public as well as investors are largely divorced from their benchmark performance and more readily felt through statements such as this from Reuters: "ChatGPT, the popular chatbot from OpenAI, is estimated to have reached 100 million monthly active users in January, just two months after launch, making it the fastest-growing consumer application in history, according to a UBS study on Wednesday" (Hu, 2023), or this from Forbes: "6 ways to make ChatGPT your productivity machine" (Cook, 2023). Rarely does anybody communicate the "success" of GPT-4 and Gemini through their SOTA performances on technical benchmarks.

Indeed, "user adoption" and "usability" have emerged as indispensable metrics in and of themselves through which technological performance is evaluated, thus somewhat shifting the locus of assessment from "technical performance" (as narrow as this definition is in the context of benchmarking) to "consumer satisfaction and adoption." As prototypes are made available as beta products, which can be seen as a development of the "agile methodology" of incremental product testing/shipping adopted by technology companies since the early 2000s (as opposed to the slower "waterfall methodology" of project management), users and consumers are effectively recruited to reinforce the projections and promises that are made around those early prototypes. Unlike the "cool demos" developed and used only by early 1950s pattern recognition researchers from specialized labs, "cool demos" today such as ElevenLabs' VoiceLab tool or OpenAI's GPT line of LLMs are not confined to a small and expert group of individuals. The lack of technical metrics is replaced by quantifiable consumer excitement, experimentation, and adoption.

On the one hand, this can be viewed as a positive development that pushes technology evaluation in a more democratic and accessible direction than the constrained task-specific technical benchmarking practices that have dominated the field of machine learning since the 1990s and definitely since the 2000s onward. On the other, however, it also more explicitly conflates financial viability and market dominance with scientific and technological progress, wherein subjective evaluations of the 3Ps ("vibe checks") disproportionately influence the direction of a technology's development and evaluation. Indeed, although it can be argued that, on a high level, technological and scientific development can never be understood separately from the vectors of capitalism, as the venture capitalist and the investment thesis fill the vacuum produced by the absence of a technical evaluative benchmark in AI development, AI systems' efficacy gets framed more explicitly through the aperture of capital, in which progress is measured as a direct function of profit. We have already witnessed the limitations of understanding not only technology development, writ large (e.g., Theranos), but also specifically artificial intelligence research this way.

As Google Deepmind CEO Hassabis suggests in his interview with *Wired*, scaling existing techniques with more compute and data may indeed yield improvements in a particular area up to a certain point, but what researchers in the field—as well as investors like Ghost—like to call "emergent qualities" are not "magically going to happen" (Knight, 2024) through such a process. Hassabis points out that the majority of machine learning techniques AI researchers are employing today, including the transformer architecture (Vaswani et al., 2017) that enables generative AI tools such as ChatGPT, Midjourney, and VoiceLab, were all invented several years ago at Google Research, Brain, and Deepmind. Indeed, despite the broader discourse permeating the tech industry and many pockets of academia that "AI is moving too fast" (Thier, 2024), which is proliferated further by the widespread availability of increasingly *usable* generative AI systems, Hassabis suggests that there have been minimal innovations (arguably close to none compared with the revolution that transformers brought to the field of machine learning in 2017) from a *scientific* perspective around how artificial intelligence systems are designed. Most, if not all, improvements have been the result of scaling existing methods. In Hassabis's words, the increased *usability* and perceived advancements are largely products of "engineering [as opposed] to research science" (Knight, 2024).

This distinction, at least in the context of this discussion, is not to downplay the role of engineering and the utility of such advances in usability. On the contrary, it only makes sense that any notion of technoscientific utility is attached firmly to a system's widespread usability by nonexperts. However, it does mean that the nature of technological progress and development is redefined primarily through the frame of industrial performance and investment potential, thereby also shifting not only the objectives of AI research and development but also the incremental steps of achieving those objectives. For instance, the widespread idea within the machine learning community that "scale is all you need" explicitly downplays the need for technological innovations and instead fastens progress and advancement to the capitalistic procurement and control of resources in the form of computing power and data. Here, the efficient management of resources around collecting, storing, and cleaning data, which also involves hiring a motley crew of human data workers (annotators, managers, cleaners, etc.) often under exploitative conditions (Perrigo, 2023), access to real estate for large-scale data centers, controlling natural resources to run and maintain those data centers, and developing strategic geopolitical and business partnerships, become key conditions for successful AI development.

Although "scientific practice" outside of explicit corporate structures is also arguably dependent on such factors, there is an undeniable difference when technological research is conducted primarily through the channel of venture capital and measured via its ability to turn a profit on speculative investments. Not only that, but the "speed of AI progress" in industry discourse is also affixed to widespread *horizontal implementation* of existing deep learning techniques to different sectors as opposed to *vertical* scientific advancements. This means that expansion of AI's "total addressable market" (TAM; and therefore the market capitalization of major AI companies such as OpenAI, Google, and NVIDIA) becomes more indicative as a measure of progress than scientific developments for which those financial markers of growth may be less apparent.[12]

As artificial intelligence increasingly becomes a domain of research and development spearheaded by both big tech and VC-backed startups, the perceived correlation between technological/scientific progress and TAM expansion will inevitably intensify. Although it cannot be said for sure whether this convergence will result in a complete departure from traditionally narrow and technical benchmarking practices, especially as deep learning techniques are applied to more explicitly "subjective" domains,[13] it is likely that the investment thesis, formulated around traditional VC factors such as the 3Ps, will increasingly serve as a central instrument (alongside or in place of traditional benchmarking practices) through which AI technologies are evaluated and therefore developed.

* * *

Any discussion of "progress" is always also a negotiation of values. It implies a positive, correct, and desirable "direction" for the domain in question, especially in the context of scientific practice. As historian and philosopher of science Thomas Kuhn (1962) writes, "To a very great extent the term 'science' is reserved for fields that do progress in obvious ways" (p. 152). But as he also argues, "Progress lies simply in the eye of the beholder" (p. 155).

Indeed, as the history of artificial intelligence research—and specifically its relationship to government and now private funding—shows, progress is not an automatic or naturally occurring quality of technoscientific ventures. It must be *enabled*. Standardized instruments, enforcement bodies, communication channels, and community conventions, among others, need to be established in order for progress to be defined, tracked, and measured.

And only once progress is deemed measurable in these ways can it be *used* as a means of garnering wider interest, attracting funding, and promoting development.

Questioning what "scientific progress" means and how it is measured in a particular scientific community is thus a way to interrogate the values that underpin it and the priorities that drive it. As I have shown in this chapter, the increasing confluence of artificial intelligence research and development and venture capital reflects a shift in the standards employed to measure progress in AI. Although it is still early to fully grasp the consequences of what these shifts will bring forth, it will be important to attend closely to the increasing influence of financial and economic frames in defining not only progress and success in AI development but also areas deemed worthy of expansion.

Conclusion: The Human Metaphor

A story of artificial intelligence is ultimately a story of metaphors. By this, I mean both in the sense of how parts of the world must be translated and represented in the data structures that underlie AI technologies, but also in the sense of AI's dependency on the metaphor as an epistemology in and of itself. As a technology founded on logics of classification and representation, artificial intelligence "valorizes some point of view and silences another. This is not inherently a bad thing—indeed it is inescapable. But it *is* an ethical choice, and as such it is dangerous—not bad, but dangerous" (Bowker & Star, 1999, p. 6).

Throughout each chapter of *Machine Voices*, I have examined both the metaphors used to represent voice as well as how the voice served as a metaphor for other "things" in the context of artificial intelligence. In chapter 1, I traced a history of the different frames and forms through which voice has been made mechanically manageable—from bodily mechanisms, through visual charts, to statistical patterns—as well as how these transformations were in service of positioning the voice as a metaphor for identity. Chapters 2 and 3 explicitly focused on AI systems designed to "listen" to the voice, in which I examined and problematized the human processes through which the voice is made mechanically legible and deployable as a proxy for auditing more intangible constructs such as human personality, capability, and emotion. And finally in chapters 4 and 5, I focused on AI systems designed to "speak" through synthetic voices, wherein human bodies, memories, and perception all had to be redescribed through the aperture of machine legible data.

To reiterate from the introduction, this book is about the relations between machine, voice, and human and how each is invoked as a metaphor for the other. Our undoubtedly real human tendency and ability to distinguish familiar voices, infer affective and physical states from vocal patterns, and even modulate the sounds of our own voices to serve different needs have all served as relatable justifications for the application of machine learning to replicate these experiences. Indeed, artificial intelligence not only depends on metaphors in its construction but also seeks to reenact them in its operation. In this concluding chapter, I attend to perhaps the most foundational metaphor of all underlying artificial intelligence: the computer as a stand-in for the human mind.

AI Before It Was AI

> The view that machines will think as man does reveals the misunderstanding of the nature of human thought.
> —Ulric Neisser, "The Imitation of Man by Machine," 1963

The phrase "artificial intelligence," which immediately prompts us to make associations between our own cognitive abilities and the properties of computational software, was not always called "artificial intelligence." In fact, Allen Newell and Herbert Simon (1956) proposed calling their logic theory machine, now considered to be one of the most important developments in early AI research, a "complex information processing system" (p. 61) or "CIP system" for short. Indeed, Newell and Simon never mention the term "artificial intelligence" in their original paper. Their only comparison to the human is to indicate that complex information processing "relies heavily on heuristic methods similar to those that have been observed in human problem solving activity" (p. 61). Outside of this initial comparison, however, Newell and Simon were relatively uninterested in suffusing their description of their logic theory machine with anthropomorphic rhetoric.

In this sense, Newell and Simon's 1956 paper is important because it brings us back to a time when the phrase "artificial intelligence" was not taken for granted.[1] Their paper is an existing artifact of a moment before the conceptual metaphor that AI currently represents to stand in for various CIP systems—which, as AI analyst Alberto Romero (2021) suggests, is an honest (albeit ugly) phrase to describe "what AI people do"—was standardized. We must remember, however, that "standardization is a process of forgetting" (Mulvin, 2021, p. 6). And in order to question or challenge that standardization, we must "recover the memory of how we got here" (p. 6). Apart from my brief historical tracing of venture capital in chapter 5, this book has largely not traced the histories of *how* with regard to the broader historical development of artificial intelligence.[2] I emphasize here that we must recover this forgotten memory of AI—before it was called "AI"—as a guiding light for how we should understand artificial intelligence and the broader network to which the banner orients us moving forward.

I am most certainly not alone in this proposition. Kevin Scott, the chief technology officer for Microsoft, one of the global corporate leaders in the development of artificial intelligence systems, has expressed explicitly that "artificial intelligence is a bad name for what it is we're doing" (Mims, 2021). Jose Hernández-Orallo (2017), a professor at the Valencian Research Institute for Artificial Intelligence, states that the field of artificial intelligence has been very successful in developing artificial systems "without featuring intelligence" (p. 398). Daron Acemoglu, an economist at the Massachusetts Institute of Technology, echoes this position that "AI is somewhat of a misnomer" (Mims, 2021).

This tempered understanding of AI as not a reflection of our own cognition but rather as simply a complex information processing system, as Newell and Simon had initially proposed, has become especially imperative today. Indeed, in June 2025, a group of Apple researchers published a paper titled "The Illusion of Thinking: Understanding the Strengths and Limitations of Reasoning Models via the Lens of Problem Complexity" (Shojaee et al., 2025). Here, the researchers question the hype around "large reasoning models" (LRMs)—that is, LLMs that generate textual explanations meant to emulate "thinking" before providing answers to prompts. Reactions to the paper were polarizing in the broader community of AI researchers. Open Philanthropy researcher Alex Lawson (2025)—in alleged "collaboration" with Anthropic's LLM Claude Opus—for instance, published a response titled "The Illusion of the Illusion of Thinking: A Comment on Shojaee et al (2025)," critiquing the experimental design of the original Apple paper to argue that the results of the experiment "do not support claims about fundamental reasoning limitations" in LRMs. In contrast, several individuals such as Gary Marcus used the opportunity to reiterate their longtime positions that the capabilities of LLMs are fundamentally different from and thus should not be conflated with human thinking processes (Marcus, 2025a, 2025b). Extended debates across social media platforms such as X and coverage from media outlets such as

Forbes (Walther, 2025) further reflected the widespread interest in thinking about current AI capabilities in direct relation to human cognition. Indeed, as cultural, economic, and geopolitical attention toward so-called artificial *general* intelligence—the ultimate vision of an apparently "humanlike" AI—transcends niche AI circles and becomes a part of the public zeitgeist, it has become especially crucial to interrogate the metaphor of human intelligence underlying the development of AI, writ large.

Artificial General Intelligence

In February of 2023, OpenAI released a manifesto titled "Planning for AGI and beyond." It states that the company's unifying "mission is to ensure that artificial general intelligence—AI systems that are generally smarter than humans—benefits all of humanity" (Altman, 2023). Of the thirty-two times that "AGI" is mentioned in this manifesto, Sam Altman, the CEO of OpenAI and the author of the piece, never provides a definition for it outside of the fact that they are meant to be "generally smarter than humans." And yet, many of our existing artificial *narrow* intelligence systems are already ostensibly "smarter"—depending on how one defines "smartness"—than *some* humans. News reports have covered cases in which ChatGPT "passed" US medical licensing exams (Kennedy, 2023), GitHub's copilot has been "writing" code since 2021, and Stable Diffusion generates "original artwork" in a matter of seconds. In fact, various recommender systems, which now seem to have become a taken-for-granted infrastructure across Internet platforms such as YouTube, Netflix, and Spotify, are considered to "know" our tastes better than some of our closest acquaintances.

There has undeniably been impressive technological progress in ANI over the past decade largely in connection with increases in computational power and the sizes of training datasets. As Fjelland (2020) writes, however, "Advances in ANI are not advances in AGI" (p. 2). The operating logic of the two (as I emphasize in chapter 2) are fundamentally different: One is bound to specificity, and the other is defined by its transcendence of it. Even with advancements in machine learning techniques—namely, deep reinforcement learning, which uses multiple layers of artificial neural networks to identify patterns and find correlations across each layer and therefore can be considered to "learn" without explicit instructions—these techniques do not escape the particularity of the task that they are learning *for*. As AI researcher

François Chollet (2019) writes, "The major successes of the field have been in building special-purpose systems capable of handling narrow, well-described tasks" (p. 7). Current machine learning applications are thus fundamentally different from the way humans learn and adapt to the world (Neisser, 1963). Returning to Newell and Simon (1956), human brains may indeed be able to process complex information. But our minds are not *only* "complex information processing" systems, nor do these similarities mean that complex information processing systems are equivalent to human brains.

What is important to consider in this constant struggle that the human brain–computer metaphor represents is the question of *which* human we are imagining as the benchmark for intelligence in AI. When Altman (2023) writes that AGI equals "AI systems that are generally smarter than humans," which humans are he referring to? The imagined human that is *artificed* in AGI is of crucial importance because generality is ironically never absolute. It is always tied to a particular context.[3]

This specter of irony—around a generality that is neither specific nor universal—has haunted conceptions of intelligence long before the advent of artificial intelligence research. Indeed, the earliest scientific efforts to define intelligence as a measurable and therefore comparable object in the human were riddled with similar tensions: "The first scientific descriptions of intelligence define it as the '*g* factor,' which stands for 'general intelligence' or 'generally ability to reason.' . . . Intelligence does not therefore refer to any particular ability but rather to all the elements of a whole . . . [Francis Galton] was trying to define intelligence as having no specific content" (Malabou, 2019/2021, pp. 21–22). The *g* factor (Spearman, 1904) and its later descendant in the intelligence quotient (IQ) are both examples of how human intelligence too has long been conceived of in the most ambiguous of ways as supposedly expressible through a single latent number derived through correlational statistical techniques—namely, factor analysis—on various test scores. It is important to note, however, that even for its most staunch proponents such as the British eugenicist Francis Galton, the practical derivation of the *g* factor in its original conceptions could not escape the specificity of particular tasks that stood in as proxies for this apparent "general" ability. As Malabou (2019/2021) writes in *Morphing Intelligence*, "But how then can a 'general' ability with no form be measured? Three main factors were assessed: visual acuity, grip strength, and reaction time" (p. 21). The reductive absurdity of these ableist vectors of measurement as representing the complexity of human intelligence needs no further words. But what is more salient for this discussion is the paradoxical *specificity* through which generality had to ultimately be measured.[4]

As Koopman (2019) writes on the emergence of both personality and intelligence as scientific objects in the early 1900s (specifically through the development of psychology as a scientific discipline), the production of selfhood along these axes of personality and intelligence was tied directly to the invention of paper-and-pencil tests—that is, the measurement tools used to "measure" personality and intelligence, without which their existence as scientifically "knowable" objects would come into question. Stephen Jay Gould's (1980) classic text, *The Mismeasure of Man*, shed light on how these tests were not only more reflective of familiarity with American culture than some form of "general intelligence" but also unable to be validated in biological terms, which ultimately resulted in some of its most faithful believers to recant their support for them.

Furthermore, despite the lure and simplistic elegance of *g*, "some researchers think that the unitary form of the *g* factor was in question from the start" (Malabou, 2019/2021, p. 125). Communication scholar Larry Gross (1974), for instance, emphasized the importance of engaging learning or the "acquisition of symbolic competence" in children along separate *modes* spanning the "lexical," "social-gestural," "logico-mathematical," and "musical," among others. In *Frames of Mind: The Theory of Multiple Intelligences*, developmental psychologist Howard Gardner (1983) built on this idea to similarly posit that at least seven different types of intelligences were present in individuals to different degrees, together comprising "intellectual personalities." Psychologist Louis L. Thurstone (1935), in his book *The Vectors of Mind*, was perhaps the earliest and most outspoken proponent of multifactorial intelligence, a frame that similarly sees aptitudes in different domains of cognitive activity as independent and not deriving from a single latent variable related to a "general" intelligence. Indeed, in critiquing the arbitrariness of Spearman's *g*, Thurstone (1940) argued: "Such a factor can always be found routinely for any set of positively correlated tests, and it means nothing more or less than the average of all the abilities called for by the battery as a whole. Consequently, it varies from one battery to another and has no fundamental psychological significance beyond the arbitrary collection of tests that anyone happens to put together" (p. 208).

Regardless of one's acceptance of intelligence (singular or multiple) as a scientifically measurable concept *at all*, what is undeniably true is that even among those individuals who sought to ground and instrumentalize it—and who were thus unified in their application of scientific methods to try and define intelligence—there was little agreement.[5] Similar to my discussion of emotion as a psychometric construct in chapter 3, there is simply no scientific consensus on what "intelligence" is, where its boundaries are, and how to measure/compare it. Altman's definition of AGI—"AI systems that are generally smarter than humans"—thus says very little (I would argue, practically nothing) from a *scientific/technical* standpoint, which is presumably the most, if not the *only*, important vector for a company such as OpenAI about what its formal understanding of "artificial general intelligence" is (let alone how and when it is "achieved").

What is intelligence? What is the form of the "generally" intelligent human? How does one measure the intelligence of this human? How does one ascertain its generality? How does one determine that a machine has surpassed this level of general intelligence? These are important questions to address before making a claim toward the creation of an AGI system defined in Altman's terms. Not only that, but more importantly, these are longstanding questions that transcend the expertise and work of machine learning engineers, venture capitalists, and "AI researchers" as they exist today in the context of companies such as OpenAI. They cut to foundational questions related to human agency, cognition, and ability that have been at the heart of the sciences and humanities since their inception. The question of general intelligence—artificial or not—is a philosophical, humanistic, and scientific question long before it is an engineering or economic one.

If we are to thus accept the consequences of this spectacular misnaming, that is to indeed engage with these "complex information processing" systems as truly "artificially intelligent" technologies—which has undeniably been the trajectory of the AI community in recent years—we should also be prepared to study, understand, and engage with these technologies through these more foundational and non-engineering frames. In my view, this reframing might be the most efficient path to underscoring the shortcomings of the admittedly exciting but ultimately unproductive human metaphor in the current AI community and the misnomer that it birthed. Although I am not entirely in agreement with Malabou's (2019/2021) overall views around the transformative impact of artificial intelligence research on our understanding of human intelligence, her statement here rings undeniably true: "Paradoxically, the more the artificial imitates the biological, the more it reveals their incommensurability" (p. 150).

"Alignment"

Another related but distinct aspect of the broader human metaphor and AGI narrative fueled by companies such as Microsoft's OpenAI and many others such as Google Deepmind, the Alignment Research Center founded by ex-OpenAI researcher Paul Christiano, and Anthropic also founded by ex-OpenAI researchers Dario and Daniela Amodei, is the focus on "alignment." Here, alignment simply refers to the idea that so-called AGI must be "aligned with human values and follow human intent" (Leike et al., 2022). The underlying assumptions as expressed by its practitioners are that AGI is inevitable and that, without proper measures in place, it poses an existential risk to humanity. Alignment research is thus quite literally predicated on what I am calling here a "SkyNet" imaginary.[6]

As I mentioned in chapter 2, however, many researchers working in the "responsible AI" space have pointed to these hypothetical AGI futures as overblown. In fact, these hypothetical discussions are seen as hindrances to focusing resources and attention to the *present* harms of existing AI/ML systems (Wong, 2023), such as their tendency to exacerbate social inequalities or the exploitative labor conditions on which the development and maintenance of these systems rely (Johnson, 2022). In this sense, these fears are indeed unproductive for channeling resources to address the present harms of artificial intelligence. But seen differently, they also provide a channel through which we can interrogate the underlying values that drive this broader project of alignment: *Whose* "values" and "intentions" is so-called AGI being aligned to?

Alignment initiatives are often accompanied by statements such as this one made by Altman (2023): "We want AGI to empower humanity to maximally flourish in the universe. We don't expect the future to be an unqualified utopia, but we want to maximize the good and minimize the bad, and for AGI to be an amplifier of humanity." The key question, of course, is whose humanity are they seeking to amplify? More critically, why are they in the position to determine what benefits all of humanity? What exactly is all of humanity? In the process of fine-tuning ChatGPT to ensure that it does not generate "toxic" content, for instance, OpenAI paid workers in Kenya between $1.32 and $2 per hour to label swathes of training data that were deemed to be toxic (Perrigo, 2023). It appears, here, that all of humanity doesn't include the very workers in the Global South employed for the development of their now "for-profit" products (O'Brien et al., 2024). As CEO Sam Altman is projected to receive an equity stake estimated at $150 billion (Hu & Cai, 2024) through the controversial restructuring of the company from a nonprofit to a for-profit one, the beneficiaries of OpenAI's commitment to "empowering humanity" can be seen as anything but equal and most definitely not universal. Timnit Gebru, former co-lead of Google's Ethical AI research team and founder of the Distributed AI Research (DAIR) institute, echoed these concerns when she asked during a talk at Stanford University, "Utopia for whom? Who is getting this utopian life that [big tech] is promising?" (Artandi, 2023).

Definitions of humanity, benefit, risk, utopia, and intent, all of which are central terms used to communicate and advance the broader project of AI, are ultimately tied to particular subjectivities and contexts: Each of these terms is variable and changes depending on who defines it. The concept of a value-aligned artificial "general" intelligence is thus inherently both a reductive and normative one, to borrow from Bowker and Star (1999) once again, that "valorizes some point of view and silences another" (p. 6). Even in the present day, we have already seen how social values embedded in artificial intelligence systems surface via facial recognition technologies that fail to detect Black faces (Buolamwini & Gebru, 2018), accent translation systems that apply machine learning techniques to "remove" the accents of call center employees to "sound more palatable to American ears" (Al-Sibai, 2022), AI acoustic gunshot detection systems that are deployed in predominantly Black neighborhoods (Kang & Hudson, 2022), and text-to-image machine learning models that replicate gender stereotypes in the visual representation of certain occupations (Rose, 2022). We have always needed and still need "value alignment" in artificial intelligence, but just not in the hypothetical and existentially dire frame that long-termist AGI researchers are conceiving it. Existing AI systems already exhibit selective "values" that pose "risks" to people today, and the explicit focus and investment toward a future "humanity" at the expense of ignoring our present one provides perhaps the clearest illustration of the present and future values of the AI community. Albeit in a different vein, I draw from the wisdom of Malabou (2019/2021) once more: "Intelligence would thus become the unfounded foundation of the origin of inequality among us all" (p. 3).

* * *

Our imaginaries around what voice is, the mechanical processes through which we manage it, and the various settings in which we make it useful together comprise the metaphors and stories I present in *Machine Voices*. Indeed, the voice, which at once feels so intimate and knowable while escaping any attempt to be bounded and stabilized, presents a lure and challenge for artificial intelligence. If one is to take anything away from this book, my hope is that it's an understanding that the technological spectacles of artificial intelligence and machine learning are ultimately the product of human processes delicately riddled with the same uncertainties, vulnerabilities, and subjectivities that the broader project of artificial intelligence has positioned itself against and which the messiness of voice so beautifully represents.

Notes

Introduction

1. The protagonist of this story, Eliza Doolittle, was the inspiration for the first "chatterbot," ELIZA, developed in the 1960s by Joseph Weizenbaum.

2. See Dockray et al. (2020) for a critical examination of contemporary machine listening technologies.

3. One arguable exception might be AI voice cloning. Even there, however, the objective is to identify general patterns across numerous recordings of an individual's speech. It is the same principle, simply applied to a more confined/local site.

Chapter 1

1. In a chapter (2026) coauthored with Juana C. Becerra, we parse how this "acoustic–linguistic divide" is also central to the construction of contemporary AI voice translation and accent modulation tools.

2. See also Rice (2012, 2015).

3. See Sterne (2003) for a more detailed history of "listening" technologies.

4. I use "voiceprinting" here instead of "voicegramming" because that was how it was referred to at this time.

5. See Kamine (1969) and Li and Mills (2019) for more detailed histories of Kersta's background and role in the rise of voiceprinting during the mid-twentieth century. See Potter et al. (1944) for the original research that Kersta built on to establish his voiceprint technique.

6. Harding (1992) also critiques such notions of objectivity as "weak objectivity" and suggests that a "strong objectivity" actively considers the situatedness of the researcher and its influence on scientific practice.

7. See separate studies, as well as Kamine (1969) and National Research Council (1979) for more detailed accounts of the different experiments.

8. See Amino et al. (2012), Greenberg et al. (2020), and Hansen and Hasan (2015) for more detailed histories of ASR after voiceprinting such as auditory-spectrographic and acoustic-phonetic approaches, as well as the NIST SRE.

9. This is an ongoing annual evaluation. The most recent one was held in December 2024 in San Juan, Puerto Rico.

10. Agnitio is a subsidiary of Nuance Communications, which was acquired by Microsoft in 2022.

11. See Kang and Hudson (2022) for a relevant case study in the context of the acoustic gunshot detection system ShotSpotter.

12. Hereafter, "recognition" refers to the broader domain encompassing both speaker verification and speaker identification; "verification" and "identification" refer specifically to the respective subtasks.

13. This will likely resonate with readers anecdotally. Voice ID has been a common feature in many security systems over the past decade. Forensic speaker identification, however, is still rarely accepted as evidence in court cases.

14. See Huh et al. (2024) for a more detailed history of the VoxCeleb dataset and the "VoxCeleb Speaker Recognition Challenges" (VoxSRC) hosted by *Interspeech* from 2019 to 2023. See Nagrani et al. (2020) for the 2020 VoxCeleb paper that was officially published in the journal *Computer Speech & Language*.

15. "Brute force" approaches in machine learning refer to a training approach in which models are trained on enormous amounts of un- or minimally processed data.

16. It is worth noting that 1:1 speaker verification in multifactor authentication security systems (e.g., banking) still primarily rely on non–deep learning approaches for their simplicity and resource efficiency.

Chapter 2

1. See Hardt and Recht (2021) for a history of early OCR research conducted by individuals such as Bledsoe, Browning, Chow, Highleyman, and Kamentsky.

2. These features can be distinguished not only in binary terms but also in terms of weights. In the words of Selfridge and Neisser (1960), "Some features are simply not subject to a reasonable dichotomy. An A very surely has a crossbar, an O very surely has not. But what about B? The most we can say is that it has more of a crossbar than O, and less than A" (p. 66).

3. Crawford and Paglen (2019) write similarly about image–text pairs in their essay "Excavating AI" that certain nouns are more "nouny" than others—that is, "apple" versus "health"—by which they mean that visually representing the noun "apple" is much clearer and less subject to interpretation than trying to visually represent the noun "health."

4. See Campolo and Schwerzmann (2023), Jaton (2021), and Sadowski (2025) for more detailed explanations on how ground truths are also relevant beyond "supervised" learning models.

5. I elaborate on this in more detail in the conclusion of this book.

6. This essentializing statement should not be thought to be at odds with research that has pointed to the hyper-adaptability of algorithms (e.g., Seaver, 2019). There, new data that is processed by an algorithmic system (e.g., a recommendation system) refines the generated outputs. Although the ability to "better" execute the task changes based on new data, the *nature* of the task (recommendation) does not change. New data is only considered in *relation* to the specific task that the algorithmic system is meant to execute, and that task is defined by the ground truth.

7. Van Rossem and Pelizza (2022) also introduce a method called the "ontology explorer" (OE) for retroactively investigating data infrastructures. Although OE takes a content analysis-type quantitative approach to foreground the systematic comparison of diverse information systems, GTT takes a more qualitative approach for discursively tracing the ways in which qualitative phenomena are translated into ML-compatible forms, thus attending more specifically to the ontological translations that emerge out of "making a problem learnable." GTT is also not the same as "Explainable AI" (Gunning et al., 2019), which emphasizes a transparency-focused approach to understanding the *algorithmic* processes that an ML system takes to generate a certain output. GTT instead focuses on the decisions and assumptions made by *humans* in constructing the ground truth for an ML system.

8. As Mackenzie (2017) writes, "Learning is comprehensively understood in machine learning as finding a mathematical *function* that could have generated the data and optimizing the search for that function as much as possible" (p. 46).

9. Star (1983) traces this process of "simplification" in scientific practice, writ large.

10. Han (2009), for instance, documents biased "policing practices that produce forms of race-like status for rural migrants in Chinese cities" (p. 593), recounting an instance in which a college student was stopped by the police for looking like a rural migrant due to his "rough skin and course flesh" (p. 602).

11. Since this time, many companies have emerged that sell similar "AI" technologies to streamline recruiting and interviewing processes. See Schellmann (2024) for an extensive overview.

12. All names of interviewees are pseudonyms unless stated otherwise.

13. In its *most basic / reduced* iteration—that is, when datasets or access to annotators/engineers are almost entirely barred—GTT can also, at the very least, provide grounds for conducting due diligence on the part of the critic/auditor helping to displace the burden of proof (of harm or bias) from the auditor to the developer as a burden of accountability—that is, based on what this ML system claims to do, this is how employee fit must be ground-truthed, and GTT shows that it's a weak and reductive approach; if this is wrong, explain why. This can thus address situations in which developers resort to "black box" arguments in which the opacity of ML approaches is leveraged to offload accountability. The most *realistic* (although not necessarily *ideal*) way in which GTT might be applied is as an amalgamation of deductive reasoning and empirical tracing, in which there is room for a shifting balance between the two, depending on how much data around the machine learning system is available.

14. IQ is an acronym for "intelligence quotient" and is one of the most heavily contested topics in psychometric research (e.g., Gould, 1980; Malabou, 2021; Stanovich, 2009). OCEAN is also an acronym, in which O stands for "openness," C for "conscientiousness," E for "extroversion," A for "agreeableness," and N for "neuroticism." The validity of OCEAN as a measure of personality has also been challenged extensively (e.g., Chun, 2021).

15. Chapter 1 examines this in the context of speaker recognition technologies.

16. This is also verifiable in a patent document that lists Adam as the inventor and the company as the assignee. I do not cite the patent here to adhere to privacy protocols.

17. Also discussed in chapter 1.

18. Bluntly stated, such individuals contribute to the bottom line by being inexpensive to screen (i.e., legibly attractive to algorithmic screening tools).

19. Real name used with consent. At the time of the interview, Rachael was working as a staff developer advocate for Rasa, where she helped developers build and deploy ML chatbots using the company's open-source framework.

20. Other professional engineers in the US adhere to a "standard of care" (Dal Pino, 2014) as part of their civil liability.

21. Serra-Garcia and Gneezy (2021) show the attraction to "unbelievable" findings is also present in academia, as they observe non-replicable publications across the sciences often have more citations than replicable ones, even after the non-replicable studies are called out for their shortcomings.

22. A startup that has a valuation of at least $1 billion.

23. I delve into these industry dynamics in more detail in chapter 5.

24. The justification for generative AI used to generate images or audio within a GTT framework might be framed around the relatively low stakes of their outcomes along with the advantages of "artistic interpretability" that allow for various outputs to be considered "acceptable" (Orr & Kang, 2024). This is not to take away from the performance of such generative models. There are indeed conceptual benchmarks for such systems as well. For instance, if the input phrase for an image generator was "a dog eating a popsicle," and the image generator produced an image of a cat licking a lollipop, it would point to inadequate labeling processes and poor overall performance. However, as long as the generated image includes *some* configuration of a dog eating a popsicle, it will be considered "acceptable." Its classification as a "creative" application means it gets to forgo the arguably tighter spectrum of "acceptability" of other ML tasks such as autonomous driving.

Chapter 3

1. From the Proceedings of the LREC 2010 Workshop on Corpora for Research on Emotion and Affect: "The VENEC corpus consists of 100 professional actors from 5 English speaking cultures (USA, India, Kenya, Singapore, and Australia) who vocally expressed 19 different affects/emotions (affection, amusement, anger, contempt, disgust, distress, fear, guilt, happiness, interest, lust, negative surprise, neutral, positive surprise, pride, relief, sadness, serenity, and shame), each with 3 levels of emotion intensity, by enacting finding themselves in various emotion-eliciting situations" (Laukka et al., 2010).

2. The theory of discrete emotions is also referred to as a "categorical" view of emotion or simply "basic" emotion theory.

3. See Ekman (1970), Izard (1994), and Tomkins (1962, 1963) for more detailed descriptions and support for "discrete emotion theory." Crawford (2021) also provides a historical overview and critique of Ekman's work in the context of facial emotion recognition technologies.

4. It should be made clear here that the ontological and epistemological complexities of vocalized emotion are not necessarily as salient in everyday contexts, in which labels such as "fear" and "anger" can and are used to characterize a vast range of events, sensations, and experiences rather smoothly. They become salient in *scientific* contexts in which these semantic emotional labels must be empirically established to bound particular configurations of physiology and behavior as particular emotions.

5. See Schuller (2018) for a more comprehensive history of SER.

6. Stark and Hoey (2018) also write in detail on the unstable paradigms of emotions used in "emotional artificial intelligence" or EAI systems (McStay, 2018).

7. Jaton (2017) suggests that guiding questions to be addressed in a "problem-oriented" approach include "How was the problem defined? How was the dataset collected? How was the crowd-sourcing experiment conducted?" (p. 833).

8. In line with the STS sensibilities of this book, I take the position that concepts need to be massaged and redefined to be made legible to science. In this way, concepts such as "emotion" are *made* scientific.

9. See chapter 1 for a more detailed discussion.

10. This is not unlike the formation of communities of scientific experimentation outlined by Kuhn (1962) or Shapin and Schaffer (1985).

11. In a widely cited paper accompanying the RAVDESS multimodal emotion dataset, Livingstone and Russo (2018) explicitly state, "While the discrete model of emotion has been criticized . . . it is a practical choice in the creation and labeling of emotion sets" (p. 6). Munn (2020) also elaborates on the *operational* value of basic/discrete emotion theory despite having a "highly impoverished understanding of emotionality" (p. 27).

12. Arousal is often used interchangeably with activation in the psychology literature around emotion.

13. Although the majority of SER training data use the categorical model of discrete emotion theory, some studies have employed the core affect model as a method of labeling affect. One notable example is in Semel's (2022) ethnography of a team developing a machine listening application that predicts whether a speaker will have a manic or depressive bipolar episode. Although Semel does not comment directly on the efficacy or relevance of using this affective model during the labeling process, she does critique the paradoxical notion of trying to listen only for acoustic affect and not linguistic content. It should be made clear here that my juxtaposition of core affect against discrete emotion theory is not necessarily to advocate for one over the other but rather to direct attention toward the unsound comparisons made by Cowen et al.'s (2019) paper in justifying the scientific robustness of a categorical view of emotion.

14. A comparable (but, of course, not equivalent) scientific tool is the Kelvin, which is a scientific unit for measuring temperature, but does not necessarily stand in to replace the experience of climate in everyday discourse over more commonly used terms such as "chilly" or "warm." For instance, one would expect the results of a study that asked participants unfamiliar with K units to describe the climate of a room via temperature-related labels versus numerically in Kelvin to show that participants are generally more likely to consistently describe a 10°C room as "chilly" than as 283 K. Most would agree that this does not mean that the descriptor "chilly" is a more scientifically useful concept than 283 K.

15. Mulvin (2021) calls this more general scientific practice "proxification."

16. As stated in footnote 13, this is one example of a machine listening technology that does not use the dominant framework of discrete emotion theory but instead tries to employ Russell's (2003) model of core affect.

17. In a chapter titled "Dissecting Speech, Manufacturing Difference: Troubling the 'Acoustic-Linguistic Divide' in Artificial Intelligence," (2026) Becerra and I explore this divide in more detail.

18. This would, of course, also be true had an emotional vocalization been recorded from what researchers call "natural" (nonexperimental or noncontrolled) settings. The point here is not to argue that one is "more authentic" than the other but rather that emotional expressions are inextricably tied to the contexts in which they emerge.

19. In the context of facial emotion recognition, Cabitza et al. (2022) show that there is an "unbearably" low reliability for facial emotion recognition ground truths, which means that there is extreme variability in people's emotion ratings of faces. Based on the results of their experiment, the authors assert that "*we cannot speak of accuracy for facial expression and emotion recognition technology*: in fact, no reference can be reliably established against which to compute meaningful error rates" (p. 14).

20. This is a shorthand used to refer to technologies that assist disabled people. It has been critiqued, however, for its redundancy—all technology is assistive, not just technology that assists disabled individuals—and paternalism. Mills (2015) has also argued that "the phrase advances a technological fix that is unconcerned with education, community support, or social change" (p. 178).

21. Google reCAPTCHA authentications are perhaps the most well-known and common iteration of reverse Turing tests encountered in everyday life.

22. An FER system developed by the recruiting-technology firm HireVue similarly analyzed facial movements to denote the potential productivity and overall "employability" of a candidate (Harwell, 2019). HireVue eventually removed this function from their recruiting technologies not only "due to the prolonged criticism that Ekman's universality thesis has faced by cultural anthropologists and others" (Bucher, 2022) but also, in part, because of the inherently unethical nature of using AI-assisted physiognomy to evaluate potential workers.

Interlude 2

1. Popular shortform for Scarlett Johansson.

2. "Testers for the nonprofit group called more than 60 insurance offices and sought information about renters' insurance. In 150 cases, responses to black and Latino callers were compared with responses to white callers, and 45 percent revealed discrimination. . . . The council's report cited a range of cases in which minorities faced both subtle and overt discrimination" (Horwitz, 1999).

3. Connor writes in a more philosophical context, but his framework of the vocalic body is presented against Altman's "sound hermeneutic," which explicitly examines the disembodied voice in film.

Chapter 4

1. One exception is Ramati's (2024) discussion of AI voices through his framework of "algorithmic ventriloquism."

2. Discussed briefly in interlude 1.

3. Short for "modification," modding in video games refers to the practice of users tweaking aspects of the original game to varying degrees, which can range from minor tweaks to the development of entirely new plots, objectives, and aesthetics.

4. "PseudoVoice AI" is a pseudonym for the company. The employee I spoke with indicated that although she wouldn't mind being identified by name herself, she would like to keep the name of the company she work(ed) for at the time as a pseudonym.

5. In Eidsheim's (2018) research, "health and "naturalness" are understood in relation to a singer's ethnicity or race.

6. This resembles the way that the actors in chapter 3 were instructed to vocalize emotions using the FEA method.

7. Tamara differentiated this technique from method acting. Although method acting is generally understood as embodying many of the aspects of "becoming" that Tamara describes, for Tamara, "becoming" constituted a deeper engagement with the character than that of method acting. It is also important to note here that actors often have different interpretations of acting methods and will adapt and interpret them to craft their own acting styles.

8. This aligns with my critique of FEA in chapter 3.

9. Jane also goes by her pen name J. A. Thomas.

10. Along this same vein, Jane expressed disappointment around the broader critique around Asian American actress and comedian Awkwafina (born Nora Lum), who has been accused of appropriating a "blaccent" or "Black Accent" (Levenson, 2022). She expressed that this public critique was hard for her to watch because it resonated with her own experience growing up as an Asian woman in Boston—"[Awkwafina's] voice is just a reflection of the fact that she grew up in Queens. Why can't she sound like that?"

11. I ended each interview by explicitly asking voice actors about their opinions on AI voice generation tools.

12. Bloomberg recently reported that ElevenLabs achieved a valuation of $1.1 billion with investments from major Silicon Valley venture capital firms such as Andreessen Horowitz and Sequoia (Bergen, 2024).

13. In May 2024, Scarlett Johansson released a public letter indicating her disapproval of OpenAI's "Sky," which employed a voice that sounded uncannily familiar to Johansson's voice, despite her refusal of the invitation to participate in the system's development (Elias, 2024).

14. In a forthcoming piece titled "The Voice Disaggregate: A Computational Ontology," I reflect on this modularity through a theoretical frame I call the "voice disaggregate."

15. As a bilingual person myself who has experimented with transforming a recording of my English speaking into an AI-generated Korean speaking one, I believe that the AI-generated Korean voice sounds qualitatively different than my embodied Korean speaking voice.

Chapter 5

1. Larry Ellison was listed as the eighth wealthiest person in the world by the Bloomberg Billionaire Index in March 2024 and the cofounder of Oracle Corporation, and Tip Draper is the founder of the venture capital firm Draper Fisher Jurvetson and was a notoriously early investor in "disruptive" technology companies such as Baidu, Hotmail, Skype, SpaceX, Tesla, and Twitter.

2. I discuss the dynamics of the Netflix competition in more detail in my paper with Orr, "AI as a Sport: On the Competitive Epistemologies of Benchmarking" (2024).

3. The film *The Big Short* (McKay, 2015), based on the book *The Big Short: Inside the Doomsday Machine* (2010) by Michael Lewis, shows the immense consequences of how an unhinged derivatives market that operates in relation to—but also separately from—mortgage loans fueled the collapse of the US economy in 2008.

4. Chapter 4 covers this topic in more detail.

5. Recall the same language used in the opening statement of the inaugural VoiceMOS challenge (Huang et al., 2022).

6. Indeed, datasets must *become* benchmarks. And there is a level of competition through which a dataset or group of datasets must reign supreme as the standardized setting for comparison.

7. Netzorg et al. (2024), for instance, develop a protocol around "perceptual voice qualities (PQs)" (p. 1) as an alternative means of interpreting speech that diverges from the MOS framework.

8. This person preferred to remain anonymous—a "ghost" in the person's words. Therefore, I am using the pseudonym Ghost.

9. Here, Ghost primarily refers to "creation" in the context of media such as films and video games, both of which, of course, are at the center of lucrative consumer industries.

10. Nicholas (2019) also writes about the "investing styles" of Silicon Valley venture capital firms listing subjective criteria such as "integrity," "motivation," and "leadership," among other qualities, in judging founders/CEOs used by various renowned venture capitalists such as Tommy K. Davis Jr. of Davis & Rock (p. 199).

11. Technical benchmarks are also scaled up subjective measures, but there is a difference in the degree to which that subjectivity is upheld as a method of evaluation compared with an explicitly subjective evaluation based on "personal experience."

12. This can be compared to the distinction between "basic science" and "applied science."

13. Fu et al. (2024) have recently attempted to benchmark "commonsense understanding" in the context of text-to-image models.

Conclusion

1. What is referred to as the computational theory of mind, which emerged alongside the cognitive revolution (from a behaviorist paradigm) in the field of psychology, is a relevant history here, but it is beyond the scope of this conclusion and book. See Miller (2003), Neisser (1967), Putnam (1988), and Rescorla (2020) for additional context.

2. See Edwards (1996), Fjelland (2020), Kline (2015), and Mendon-Plasek (2021) for more detailed histories around cybernetics, artificial intelligence, and machine learning.

3. As Chollet (2019) also writes, even ambiguous conceptions of AGI are primarily bound to human generality as opposed to a truly universal generality.

4. There have been subsequent developments that have reclaimed the *g* factor through more comprehensive tests (compared with the original evaluations related to visual acuity, grip strength, and reaction time), and there is a general consensus on its reliability as a statistical regularity. The source of the positive correlations underlying *g*, however, are still widely contested.

5. This has been one of the most highly debated topics in psychological research. Gould (1980) and Kamin (1974) provide technical, social, and political critiques of psychometric approaches. Haier (2016) provides an overview of the neuroscientific support for *g*. Herrnstein and Murray (1994) examine the alleged predictive validity

of *g*, including across racial groups, which resulted in the highly contested and controversial work *The Bell Curve*. Stanovich (2009) argues that despite broader claims, *g* is a measure of a specific type of narrow intelligence that doesn't adequately capture what we meaningfully refer to as intelligence in the everyday.

6. In the *Terminator* films, SkyNet "is a highly advanced computer system possessing artificial intelligence. Once it became self-aware, it saw humanity as a threat to its existence" (Terminator Wiki, n.d.).

References

a16z. (2024). *The Abundance Agenda.* https://gamma.app/public/a16z-Consumer-Abundance-Agenda-ieotbnzbxj81biu?mode=doc

Acharya, A. (2024). *How AI will usher in an era of abundance.* a16z. https://a16z.com/how-ai-will-usher-in-an-era-of-abundance/

Akhtar, S. (2012). *Psychoanalytic listening: Methods, limits, and innovations.* Karnac Books.

Al-Sibai, N. (2022, August 23). *Cursed startup using AI to remove call center workers' accents.* https://futurism.com/startup-ai-remove-call-center-accents

Alaa, A., Hartvigsen, T., Golchini, N., Dutta, S., Dean, F., Raji, I. D., & Zack, T. (2025). *Medical large language model benchmarks should prioritize construct validity.* ArXiv. https://doi.org/10.48550/arXiv.2503.10694

Alper, M. (2018). *Can technology really "give voice" to disabled people?* https://psmag.com/social-justice/can-technology-really-give-voice-to-disabled-people

Altman, R. (1992). *Sound theory / Sound practice.* Routledge.

Altman, S. (2023). *Planning for AGI and beyond.* https://openai.com/blog/planningfor-agi-and-beyond

Ames, M. G. (2019). *The charisma machine: The life, death, and legacy of One Laptop per Child.* MIT Press.

Amino, K., Osanai, T., Kamada, T., Makinae, H., & Arai, T. (2012). Historical and procedural overview of forensic speaker recognition as a science. In A. Neustein & H. A. Patil (Eds.), *Forensic speaker recognition* (pp. 3–20). Springer.

Ananny, M., & Crawford, K. (2016). Seeing without knowing: Limitations of the transparency ideal and its application to algorithmic accountability. *New Media & Society, 20*(3). https://doi.org/10.1177/1461444816676645

Anderson, W. S. (1905). Laryngology. *The Physician and the Surgeon: A Professional Medical Journal, 27*(1), 228.

Andreessen, M. (2023a). *The techno-optimist manifesto.* a16z. https://a16z.com/the-techno-optimist-manifesto/

Andreessen, M. (2023b). *Why AI will save the world.* a16z. https://a16z.com/ai-will-save-the-world/

Andrews, M. (2024). *The immortal science of ML: Machine learning and the theory-free ideal.* https://www.researchgate.net/profile/Mel-Andrews/publication/371982028_The_Immortal_Science_of_ML_Machine_Learning_the_Theory-Free_Ideal/links/66d77129fa5e11512c7336ec/The-Immortal-Science-of-ML-Machine-Learning-the-Theory-Free-Ideal.pdf

Aratani, L. (2025, May 30). FBI investigating efforts to impersonate White House chief of staff Susie Wiles. *The Guardian.* https://www.theguardian.com/us-news/2025/may/30/susie-wiles-trump-chief-of-staff-impersonation

Artandi, S. (2023). "Utopia for whom?": Timnit Gebru on the dangers of artificial general intelligence. *The Stanford Daily.* https://stanforddaily.com/2023/02/15/utopia-for-whom-timnit-gebru-on-the-dangers-of-artificial-general-intelligence/

Bai, Z., & Zhang, X. (2021). Speaker recognition based on deep learning: An overview. *Neural Networks, 140,* 65–99. https://doi.org/10.1016/j.neunet.2021.03.004

Baio, A. (2022). *AI data laundering: How academic and nonprofit researchers shield tech companies from accountability.* https://waxy.org/2022/09/ai-data-laundering-how-academic-and-nonprofit-researchers-shield-tech-companies-from-accountability/

Bajarin, T. (2022). *2022—The year that AI took center stage.* Forbes. https://www.forbes.com/sites/timbajarin/2022/12/28/2022the-year-ai-took-center-stage/?sh=5ef857b87f03

Bänziger, T., Mortillaro, M., & Scherer, K. R. (2012). Introducing the Geneva Multimodal expression corpus for experimental research on emotion perception. *Emotion, 12*(5), 1161–1179.

Barrett, L. F. (2017). The theory of constructed emotion: An active inference account of interoception and categorization. *Social Cognitive and Affective Neuroscience, 12*(1), 1–23.

Barthes, R. (1981). *The grain of the voice: Interviews 1962–1980.* Northwestern University Press.

Bates, J. (1994). The role of emotion in believable agents. *Communications of the ACM, 37*(7), 122–125.

"BATVOX Advanced School." (2012). *Agnitio.* https://www.scribd.com/document/102858076/Batvox-Advanced-School-Brochure-June-2012

Belin, P., Fillion-Bilodeau, S., & Gosselin, F. (2008). The Montreal Affective Voices: A validated set of nonverbal affect bursts for research on auditory affective processing. *Behavior Research Methods, 40*(2), 531–539.

Bender, E. M., Gebru, T., McMillan-Major, A., & Mitchell, M. (2021). On the dangers of stochastic parrots: Can language models be too big? In *Proceedings of the 2021 ACM Conference on Fairness, Accountability, and Transparency* (pp. 610–623). Association for Computing Machinery. https://doi.org/10.1145/3442188.3445922

Bergen, M. (2024). *AI voice-cloning startup ElevenLabs valued at $1.1 billion*. Bloomberg. https://www.bloomberg.com/news/articles/2024-01-22/ai-voice-cloning-startup-elevenlabs-valued-at-1-1-billion

Bergson, H. (1889/2001). *Time and free will: An essay on the immediate data of consciousness*. Dover Publications.

Bernhardsson, E. (2017). *Language pitch*. https://erikbern.com/2017/02/01/language-pitch.html

Bird, S. (2022). *Responsible AI investments and safeguards for facial recognition*. https://azure.microsoft.com/en-us/blog/responsible-ai-investments-and-safeguards-for-facial-recognition/

Birhane, A., & Prabhu, V. U. (2021). *Large image datasets: A pyrrhic win for computer vision?* 2021 IEEE Winter Conference on Applications of Computer Vision (WACV), Waikoloa, HI, United States. https://doi.org/10.1109/WACV48630.2021.00158

Black, J. W., Lashbrook, W., Nash, E., Oyer, H. I., Pedrey, C., Tosi, O. I., & Truby, H. (1973). Reply to "Speaker identification by speech spectrograms: some further observations" [R. H. Bolt, F. S. Cooper, E. E. David, P. B. Denes, J. M. Pickett, and K. N. Stevens, J. Acoust. Soc. Am. 54, 531–534 (1973)]. *The Journal of the Acoustical Society of America, 54*(2), 535–537. https://doi.org/10.1121/1.1913614

Bojanic, M., Delic, V., & Karpoy, A. (2020). Call redistribution for a call center based on speech emotion recognition. *Applied Sciences, 10*(13), 4653.

Bolt, R. H., Cooper, F. S., David, E. E., Jr., Denes, P. B., Pickett, J. M., & Stevens, K. N. (1973). Speaker identification by speech spectrograms: Some further observations. *The Journal of the Acoustical Society of America, 54*(2), 531–534. https://doi.org/10.1121/1.1913613

Bowker, G. C., & Star, S. L. (1999). *Sorting things out: Classification and its consequences*. MIT Press.

Boyd, K., & Andalibi, N. (2023). Automated emotion recognition in the workplace: How proposed technologies reveal potential futures of work. In *ACM Conference on Computer Supported Cooperative Work (CSCW '23)* (Vol. 7, pp. 1–37). Association for Computing Machinery.

Bregel, S. (2024). *Here's where Ivy league colleges stand on SATs, ACTs, and standardized testing, post-pandemic*. Fast. https://www.fastcompany.com/91034996/ivy-league-colleges-sat-act-standardized-test-application-requirements-post-pandemic

Bryan, D., & Rafferty, M. (2006). *Capitalism with derivatives: A political economy of financial derivatives, capital and class*. Palgrave Macmillan.

Bucher, T. (2022). Facing AI: Conceptualizing "fAIce communication" as the modus operandi of facial recognition systems. *Media, Culture & Society, 44*(4), 638–654.

Buolamwini, J., & Gebru, T. (2018). Gender shades: Intersectional accuracy disparities in commercial gender classification. In *Proceedings of the 2018 ACM Conference Proceedings on Fairness, Accountability and Transparency* (pp. 77–91). Association for Computing Machinery.

Burma, J. H. (1946). The measurement of negro passing. *American Journal of Sociology, 52*(1), 18–22.

Burrell, J. (2016). How the machine "thinks": Understanding opacity in machine learning algorithms. *Big Data & Society, 3*(1), 1–12.

Busso, C., Bulut, M., Lee, C. C., Kazemzadeh, A., Mower, E., Kim, S., Chang, J. N., Lee, S., & Narayanan, S. S. (2008). IEMOCAP: Interactive emotional dyadic motion capture database. *Language Resources and Evaluation, 42*(4), 335–359.

Busso, C., Parthasarathy, S., Burmania, A., AbdelWahab, M., Sadoughi, N., & Mower Provost, E. (2016). MSP-IMPROV: An acted corpus of dyadic interactions to study emotion perception. *IEEE Transactions on Affective Computing, 8*(1), 67–80.

Cabitza, F., Campagner, A., & Mattioli, M. (2022). The unbearable (technical) unreliability of automated facial emotion recognition. *Big Data & Society, 9*(2).

Campolo, A., & Schwerzmann, K. (2023). From rules to examples: Machine learning's type of authority. *Big Data & Society, 10*(2), 20539517231188725. https://doi.org/10.1177/20539517231188725.

Cao, H., Cooper, D. G., Keutmann, M. K., Gur, R. C., Nenkoya, A., & Verma, R. (2014). CREMA-D: Crowd-sourced emotional multimodal actors dataset. *IEEE Transactions on Affective Computing, 5*(4), 377–390.

Carter, S. L. (2024). *The Ivy league is right to revive the SAT*. Bloomberg. https://www.bloomberg.com/opinion/articles/2024-03-08/the-sat-is-back-at-mit-brown-dartmouth-other-top-colleges-take-note?embedded-checkout=true

Catanzaro, M., Tola, E., Hummel, P., & Viciano, A. (2017). *Voice analysis should be used with caution in court*. https://www.scientificamerican.com/article/voice-analysis-should-be-used-with-caution-in-court/

Cavarero, A. (2005). *For more than one voice: Toward a philosophy of vocal expression*. Stanford University Press.

Ceraso, S. (2022). *Voice as ecology: Voice donation, materiality, identity*. Sounding Out! https://soundstudiesblog.com/2022/09/06/voice-as-ecology-voice-donation-materiality-identity/

Cheney-Lippold, J. (2017). *We are data: Algorithms and the making of our digital selves.* New York University Press.

Children's Health. (n.d.). *When do babies start talking?* https://www.childrens.com/health-wellness/when-do-babies-start-talking

Chion, M. (1999). *The voice in cinema.* Columbia University Press. (Original work published 1982)

Chollet, F. (2019). *On the measure of intelligence.* ArXiv. https://doi.org/10.48550/arXiv.1911.01547

Christin, A. (2020). The ethnographer and the algorithm: beyond the black box. *Theory and Society, 49*(5), 897–918.

Chun, W. (2021). *Discriminating data: Correlation, neighborhoods, and the new politics of recognition.* MIT Press.

"Cogito." (n.d.). https://cogitocorp.com/

Coltin, J. (2023). *Greetings from mayor Adams, generated by AI, in different languages.* Politico. https://www.politico.com/news/2023/10/16/nyc-adams-ai-languages-00121744

Confino, P. (2024). *Another AI unicorn? $80 million series B led by Andreessen Horowitz yields a $1.1 billion valuation, source says.* Fortune. https://fortune.com/2024/01/22/ai-unicorn-andreesen-horowitz-eleven-labs-venture-capital-series-b-tech/

Connor, S. (2000). *Dumbstruck: A cultural history of ventriloquism.* Oxford University Press.

Cook, J. (2023). *6 ways to make ChatGPT your productivity machine.* Forbes. https://www.forbes.com/sites/jodiecook/2023/07/17/6-ways-to-make-chatgpt-your-productivity-machine/

Cowen, A. S., Laukka, P., Elfenbein, H. A., Liu, R., & Keltner, D. (2019). The primacy of categories in the recognition of 12 emotions in speech prosody across two cultures. *Nature Human Behaviour, 3*(4), 369–382.

Crawford, K. (2021). *The atlas of AI: Power, politics and the planetary costs of artificial intelligence.* Yale University Press.

Crawford, K., & Paglen, T. (2019). *Excavating AI: The politics of training sets for machine learning.* The AI Now Institute. https://excavating.ai/

D'Cruze, D. (2023). *ChatGPT is a "horrible product": OpenAI CEO Sam Altman.* Business Today. https://www.businesstoday.in/technology/news/story/chatgpt-is-a-horrible-product-openai-ceo-sam-altman-370284-2023-02-15

Dal Pino, J. (2014). *Do you know the standard of care?* White paper. American Council of Engineering Companies.

Darwin, C. (1872). *The expression of emotions in animals and man.* Murray.

Daston, L., & Galison, P. (1992). The image of objectivity. *Representations, 40*, 81–128. https://doi.org/10.2307/2928741

Daston, L., & Galison, P. (2007). *Objectivity*. Zone Books.

Data Monsters. (2023). *Real-time accent correction AI*. https://www.datamonsters.com/technologies/accent-correction

Davis, L. E., Gallman, R. E., & Gleiter, K. (1997). *In pursuit of Leviathan: Technology, institutions, productivity, and profits in American whaling, 1816–1906*. University of Chicago Press.

Dellaert, F., Polzin, T., & Waibel, A. (1996). Recognizing emotion in speech. In *Proceeding of Fourth International Conference on Spoken Language Processing. ICSLP '96* (Vol. 3, 1970–1973). IEEE.

Deng, J., Dong, W., Socher, R., Li, L., Li, K., & Fei-Fei, L. (2009). ImageNet: A large-scale hierarchical image database. In *2009 IEEE Conference on Computer Vision and Pattern Recognition* (pp. 248–255). Institute of Electrical and Electronics Engineers. https://doi.org/10.1109/CVPR.2009.5206848

Denton, E., Hanna, A., Amironesei, R., Smart, A., & Nicole, H. (2021). On the genealogy of machine learning datasets: A critical history of ImageNet. *Big Data & Society, 8*(2), 1–14. https://doi.org/10.1177/20539517211035955

Desrosières, A. (1998). *The politics of large numbers: A history of statistical reasoning* (C. Naish, Trans.). Harvard University Press. (Original work published 1993)

Dhaliwal, R., Lepage-Richer, T., & Suchman, L. (2024). *Neural networks*. University of Minnesota Press.

Dockray, S., Parker, J., & Stern, J. (2020). *Machine listening*. https://machinelistening.exposed/

Dolar, M. (2006). *A voice and nothing more*. MIT Press.

Domingos, P. (2012). A few useful things to know about machine learning. *Communications of the ACM, 55*(10), 78–87. https://doi.org/10.1145/2347736.2347755

Dror, O. E. (2001). Counting the affects: Discoursing in numbers. *Social Research, 68*(2), 357–378.

Edwards, P. N. (1996). *The closed world: Computers and the politics of discourse in Cold War America*. MIT Press.

Eidsheim, N. S. (2015). Race and the aesthetics of vocal timbre. In O. Bloechl, M. Lowe, & J. Kallberg (Eds.), *Rethinking difference in music scholarship* (pp. 338–365). Cambridge University Press.

Eidsheim, N. S. (2018). *The race of sound: Listening, timbre, and vocality in African American music*. Duke University Press.

Eidsheim, N. S., & Meizel, K. (2019). Introduction: Voice studies now. In N. S. Eidsheim and K. Meizel (Eds.), *The Oxford handbook of voice studies* (pp. xiii–xli). Oxford University Press.

Ekman, P. (1970). Universal facial expressions of emotions. *California Mental Health Research Digest, 8*(4), 151–158.

Ekman, P., & Friesen, W. V. (1978). *Facial action coding system: A technique for the measurement of facial movement.* Consulting Psychologists Press.

El Kaliouby, R., & Robinson, P. (2005). The emotional hearing aid: An assistive tool for children with Asperger syndrome. *Universal Access in the Information Society, 4,* 121–134.

Elias, J. (2024). *Scarlett Johansson says OpenAI ripped off her voice after she told the company not to use it in software.* CNBC. https://www.cnbc.com/2024/05/20/scarlett-johansson-says-openai-ripped-off-her-voice-.html#:~:text=comforting%20to%20people.-,After%20much%20consideration%20and%20for%20personal%20reasons%2C%20I%20declined%20the,intimate%20relationship%20with%20a%20human.&text=As%20a%20result%20of%20their,that%20individual%20rights%20are%20protected.%22

Elish, M. C. (2019). Moral crumple zones: Cautionary tales in human–robot interaction. *Engaging Science, Technology, and Society, 5,* 40–60.

Elish, M. C., & boyd, d. (2017). Situating methods in the magic of big data and AI. *Communication Monographs, 85*(1), 57–80. https://doi.org/10.1080/03637751.2017.1375130

Elliott, V. (2021). *Training self-driving cars for $1 an hour.* https://restofworld.org/2021/self-driving-cars-outsourcing/

"Emotion AI." (n.d.). https://cogitocorp.com/emotion-ai/ [Screenshot]. Copy in possession of Edward B. Kang.

"Emotion-Meter for Use at Film Previews." (n.d.). Source unknown, attached to letter Leonarde Keeler, December 5, 1946. Drawer/Unmarked/Misc. 4. Department of Defense Polygraph Institute.

Engdahl, I. (2024). Agreements "in the wild": Standards and alignment in machine learning benchmark construction. *Big Data & Society, 11*(2). https://doi.org/10.1177/20539517241242457

Eriksson, M., Purificato, E., Noroozian, A., Vinagre, J., Chaslot, G., Gomez, E., & Fernandez-Llorca, D. (2025). *Can we trust AI benchmarks? An interdisciplinary review of current issues in AI evaluation.* ArXiv. https://doi.org/10.48550/arXiv.2502.06559

Faundez-Zanuy, M., & Monte-Moreno, E. (2005). State-of-the-art in speaker recognition. *IEEE Aerospace and Electronic Systems Magazine, 20*(5), 7–12. https://doi.org/10.1109/MAES.2005.1432568

Fei-Fei, L. (2017). *ImageNet: Where have we gone? Where are we going?* [Video]. ACM Learning Center Site. https://learning.acm.org/techtalks/ImageNet

Fink, C. (2024). *ElevenLabs secures $80 million for voice cloning, AI photo app Artisse bags $6.7 million, AI Art films*. Forbes. https://www.forbes.com/sites/charliefink/2024/01/25/eleven-labs-secures-80-million-for-voice-cloning-ai-photo-app-artisse-bags-67-million-ai-art-films/

Fjelland, R. (2020). Why general artificial intelligence will not be realized. *Humanities and Social Sciences Communications, 7*(10).

Friginal, E. (2009). Threats to the sustainability of the outsourced call center industry in the Philippines: Implications for language policy. *Language Policy, 8*, 51–68.

Foucault, M. (1994). *The order of things: An archaeology of the human sciences* (A. Sheridan, Trans.). Vintage. (Original work published 1966)

Fu, X., He, M., Lu, Y., Wang, W. Y., & Roth, D. (2024). *Commonsense-T2I challenge: Can text-to-image generation models understand commonsense?* ArXiv. https://doi.org/10.48550/arXiv.2406.07546

Gardner, H. (1983). *Frames of mind: The theory of multiple intelligences*. Basic Books.

Garland, A. (Director). (2014). *Ex Machina* [Film]. Film4 & DNA Films.

Gemini Team. (2023). *Gemini: A family of highly capable multimodal models*. ArXiv. https://doi.org/10.48550/arXiv.2312.11805

Gendron, M., Roberson, D., & Barrett, L. F. (2015). Cultural variation in emotion perception is real: A response to Sauter, Eisner, Ekman, and Scott. *Psychological Science, 26*(3), 357–359.

Gignac, G. E., & Szodorai, E. T. (2024). Defining intelligence: Bridging the gap between human and artificial perspectives. *Intelligence, 104*. https://doi.org/10.1016/j.intell.2024.101832

Gilbert, R. Y. (1911, July 16). Vocal fingerprints. *San Francisco Sunday Call 110*(46), 25.

Gillespie, T. (2018). *Custodians of the Internet: Platforms, content moderation, and the hidden decisions that shape social media*. Yale University Press.

Gordon, B. (2023). *Voice machines: The castrato, the cat piano, and other strange sounds*. University of Chicago Press.

Gould, S. J. (1980). *The mismeasure of man*. W. W. Norton & Company.

Gray, M. L., & Suri, S. (2019). *Ghost work: How to stop Silicon Valley from building a new global underclass*. Harper Business.

Greenberg, C. S., Mason, L. P., Sadjadi, S. O., & Reynolds, D. A. (2020). Two decades of speaker recognition evaluation at the National Institute of Standards and

Technology. *Computer Speech & Language, 60*, 1–10. https://doi.org/10.1016/j.csl.2019.101032

Gross, L. (1974). Modes of communication and the acquisition of symbolic competence. *Teachers College Record, 75*(5), 56–80. https://doi.org/10.1177/016146817407500503

Gunning, D., Stefik, M., Choi, J., Miller, T., Stumpf, S., & Yang, G. Z. (2019). XAI explainable artificial intelligence. *Science Robotics, 4*(37), eaay7120.

Hacking, I. (1981). *Scientific revolutions*. Oxford University Press.

Haier, R. (2016). *The neuroscience of intelligence*. Cambridge University Press.

Han, D. (2009). Policing and racialization of rural migrant workers in Chinese cities. *Ethnic and Racial Studies, 33*(4), 593–610.

Hansen, J. H. L., & Hasan, T. (2015). Speaker recognition by machines and humans: A tutorial review. *IEEE Signal Processing Magazine, 32*(6), 75–99. https://doi.org/10.1109/MSP.2015.2462851

Hao, K. (2022). *Artificial intelligence is creating a new colonial world order*. https://www.technologyreview.com/2022/04/19/1049592/artificial-intelligence-colonialism/

Hao, K., & Paola Hernández, A. (2022). *How the AI industry profits from catastrophe*. https://www.technologyreview.com/2022/04/20/1050392/ai-industry-appen-scale-data-labels/

Haraway, D. (2016). *Staying with the trouble: Making kin in the Chthulucene*. Duke University Press.

Harding, S. (1992). After the neutrality ideal: Science, politics, and "strong objectivity." *Social Research, 59*(3), 567–587. http://www.jstor.org/stable/40970706

Hardt, M., & Recht, B. (2021). *The saga of Highleyman's data*. http://www.argmin.net/2021/10/20/highleyman/

Harwell, D. (2019, November 6). A face-scanning algorithm increasingly decides whether you deserve the job. *The Washington Post*. https://www.washingtonpost.com/technology/2019/10/22/ai-hiring-face-scanning-algorithm-increasingly-decides-whether-you-deserve-job/

Hayashi, T., Yamamoto, R., Inoue, K., Yoshimura, T., Watanabe, S., Toda, T., Takeda, K., Zhang, Y., & Tan, X. (2020). ESPnet-TTS: Unified, reproducible, and integratable open source end-to-end text-to-speech toolkit. In *ICASSP 2020–2020 IEEE International Conference on Acoustics, Speech and Signal Processing (ICASSP)* (pp. 7654–7658). Institute of Electrical and Electronics Engineers. https://doi.org/10.1109/ICASSP40776.2020.9053512

Hazen, B. (1973). Effects of differing phonetic contexts on spectrographic speaker identification. *The Journal of the Acoustical Society of America, 54*(3), 650–660. https://doi.org/10.1121/1.1913645

Henderson, A. (2001). *Is your money where your mouth is: Hiring managers' attitudes toward African-American vernacular English* (Publication No. AAI3003635) [Doctoral dissertation, University of Pennsylvania]. ProQuest Dissertations and Theses Global.

Henrique da Cruz Martinho, D., Wanderley Lopes, L., Dornelas, R., & Carolina Constantini, A. (2024). Can acoustic measurements predict gender perception in the voice. *PLoS One, 19*(11), e0310794. https://doi.org/10.1371/journal.pone.0310794

Hernández-Orallo, J. (2017). Evaluation in artificial intelligence: From task-oriented to ability oriented measurement. *Artificial Intelligence Review, 48*, 397–447. https://doi.org/10.1007/s10462-016-9505-7

Herrnstein, R. J., & Murray, C. (1994). *The bell curve: Intelligence and class structure in American life.* Free Press.

"The Hidden Truth." Background Material, Joliet State Penitentiary, folder/Source Materials/I/B, Department of Defense Polygraph Institute.

Hollien, H. (1974). Peculiar case of "voiceprints." *The Journal of the Acoustical Society of America, 56*(1), 210–213. https://doi.org/10.1121/1.1903257

Horwitz, S. (1999, September 29). Minority renters see insurance bias. *The Washington Post.* https://www.washingtonpost.com/wp-srv/WPcap/1999-09/29/074r-092999-idx.html

Hu, H., Xu, M. X., & Wu, W. (2007). GMM supervector based SVM with spectral features for speech emotion recognition. In *2007 IEEE International Conference on Acoustics, Speech and Signal Processing—ICASSP '07* (Vol. 4, pp. IV–413). IEEE.

Hu, K. (2023). *ChatGPT sets record for fastest-growing user base—Analyst note.* Reuters. https://www.reuters.com/technology/chatgpt-sets-record-fastest-growing-user-base-analyst-note-2023-02-01/

Hu, K., & Cai, K. (2024). *Exclusive: OpenAI to remove non-profit control and give Sam Altman equity.* Reuters. https://www.reuters.com/technology/artificial-intelligence/openai-remove-non-profit-control-give-sam-altman-equity-sources-say-2024-09-25/

Huang, W., Cooper, E., Tsao, Y., Wang, H., Toda, T., & Yamagishi, J. (2022). *The VoiceMOS Challenge 2022.* ArXiv. https://doi.org/10.48550/arXiv.2203.11389

Huh, J., Chung, J. S., Nagrani, A., Brown, A., Jung, J., Garcia-Romero, D., & Zisserman, A. (2024). The VoxCeleb speaker recognition challenge: A retrospective. *IEEE/ACM Transactions on Audio, Speech, and Language Processing, 32*, 3850–3866. https://doi.org/10.1109/TASLP.2024.3444456

"Introducing Affectiva's Emotion Recognition Through Speech." (2017). https://blog.affectiva.com/introducing-affectivas-emotion-recognition-through-speech

Izaguirre, A. (2023). Can New York's mayor speak Mandarin? No, but with AI he's making robocalls in multiple languages. *Los Angeles Times.* https://www.latimes.com

/world-nation/story/2023-10-17/can-new-yorks-mayor-speak-mandarin-no-but-with-ai-hes-making-robocalls-in-multiple-languages

Izard, C. E. (1994). Innate and universal facial expressions: Evidence from developmental and cross-cultural research. *Psychological Bulletin, 115*(2), 288–299.

James, W. (1884). What is an emotion? *Mind, 34*, 188–205.

Jaton, F. (2017). We get the algorithms of our ground truths: Designing referential databases in digital image processing. *Social Studies of Science, 47*(6), 811–840. https://doi.org/10.1177/0306312717730428

Jaton, F. (2021). *The constitution of algorithms: Ground-truthing, programming, formulating.* MIT Press.

Johnson, K. (2022). LaMDA and the sentient AI trap. *Wired.* https://www.wired.com/story/lamda-sentient-ai-bias-google-blake-lemoine/amp

Jonze, S. (Director). (2013). *Her* [Film]. Annapurna Pictures.

Jumper, J., Evans, R., Pritzel, A., Green, T., Figurnov, M., Ronneberger, O., Tunyasuvunakool, K., Bates, R., Zidek, A., Potapenko, A., Bridgland, A., Meyer, C., Kohl, S. A. A., Ballard, A. J., Cowie, A., Romera-Paredes, B., Nikolov, S., Jain, R., Adler, J., . . . Hassabis, D. (2021). Highly accurate protein structure prediction with AlphaFold. *Nature, 596*, 583–589.

Kaluszynski, M. (2001). Republican identity: Bertillonage as government technique. In J. Caplan, & J. Torpey (Eds.), *Documenting individual identity: The development of state practices in the modern world* (pp. 291–310). Princeton University Press.

Kamin, L. (1974). *The science and politics of I.Q.* Lawrence Earlbaum Associates.

Kamine, B. (1969). Voiceprint technique: Its structure and reliability. *The San Diego Law Review, 6*(2), 213–242.

Kanade, V. (2022). *What is narrow artificial intelligence (AI)? Definition, challenges, and best practices for 2022.* https://www.toolbox.com/tech/artificial-intelligence/articles/what-is-narrow-ai/

Kang, E. B. (forthcoming). The voice disaggregate: A computational ontology. In L. Coulthard, C. L. Montgomery, & K. Spring (Eds.), *The Oxford handbook of media and vocality*. Oxford University Press.

Kang, E. B., & Becerra, J. C. (2026). Dissecting speech, manufacturing difference: Troubling the "acoustic linguistic divide" in artificial intelligence. In L. Koycheva, A. VandenBroek, & M. Artz (Eds.), *Anthropology and AI* (pp. 122–135). Routledge.

Kang, E. B., & Hudson, S. (2022/2024). Audible crime scenes: ShotSpotter as diagnostic, policing, and space-making infrastructure. *Science, Technology, and Human-Values, 49*(3), 646–672. https://doi.org/10.1177/01622439221143217

Kennedy, S. (2023). *ChatGPT passes US medical licensing exam without clinician input.* https://healthitanalytics.com/news/chatgpt-passes-us-medical-licensing-exam-without-clinician-input

Kenton, W. (2023). *Investment thesis: An argument in support of investing decisions.* Investopedia. https://www.investopedia.com/terms/i/investment-thesis.asp

Kersta, L. G. (1962). Voiceprint identification. *Nature, 196*(4861), 1253–1257.

King, S., & Karaiskos, V. (2010). *The blizzard challenge 2010.* http://festvox.org/blizzard/bc2010/summary_Blizzard2010.pdf

King, S., & Karaiskos, V. (2011). *The blizzard challenge 2011.* http://festvox.org/blizzard/bc2011/summary_Blizzard2011.pdf

King, S., & Karaiskos, V. (2013). *The blizzard challenge 2013.* http://festvox.org/blizzard/bc2013/summary_Blizzard2013.pdf

King, S., & Karaiskos, V. (2016). *The blizzard challenge 2016.* http://festvox.org/blizzard/bc2016/blizzard2016_overview_paper.pdf

Kline, R. (2015). *The cybernetics moment: Or why we call our age the information age.* Johns Hopkins University Press.

Knight, W. (2024). Google's AI boss says scale only gets you so far. *Wired.* https://www.wired.com/story/deepmind-ceo-demis-hassabis-interview-artificial-intelligence-scale/

Kofman, A. (2018a). *Forget about Siri and Alexa—When it comes to voice identification, the "NSA reigns supreme."* https://theintercept.com/2018/01/19/voice-recognition-technology-nsa/?comments=1#comments

Kofman, A. (2018b). *Interpol rolls out international voice identification database using samples from 192 law enforcement agencies.* https://theintercept.com/2018/06/25/interpol-voice-identification-database/

Koopman, C. (2019). *How we became our data: A genealogy of the informational person.* University of Chicago Press.

Krämer, S. (2009). "Epistemology of the line": Reflections on the diagrammatical mind. Unpublished manuscript.

Kreiman, J., & Sidtis, D. (2011). *Foundations of voice studies: An interdisciplinary approach to voice production and perception.* John Wiley.

Kubrick, S. (Director). (1968). *2001: A space odyssey* [Film]. Stanley Kubrick Productions.

Kuhn, T. (1962). *The structure of scientific revolutions.* University of Chicago Press.

Labov, W. (1973). *Language in the inner city: Studies in the Black English vernacular.* University of Pennsylvania Press.

Laukka, P., Elfenbein, H. A., Chui, W., Thingujam, N. S., Iraki, F. K., Rockstuhl, T., & Althoff, J. (2010). Presenting the VENEC corpus: Development of a cross-cultural corpus of vocal emotion expressions and a novel method of annotating emotion appraisals. In *Proceedings of the LREC 2010 Workshop on Corpora for Research on Emotion and Affect* (pp. 53–57). European Languages Resources Association.

Lawson, A. (2025). *The illusion of the illusion of thinking: A comment on Shojaee et al (2025)*. ArXiv. https://arxiv.org/html/2506.09250v1

Leike, J., Schulman, J., & Wu, J. (2022). *Our approach to alignment research*. https://openai.com/blog/our-approach-to-alignment-research

Leshin, J., Carter, M. J., Doyle, C. M., & Lindquist, K. A. (2024). Language access differentially alters functional connectivity during emotion perception across cultures. *Frontiers in Psychology, 14*, 1084059. https://doi.org/10.3389/fpsyg.2023.1084059

Levenson, E. (2022). Awkwafina issues statement addressing accusations that she has used a blaccent.' CNN. https://edition.cnn.com/2022/02/06/entertainment/awkwafina-blaccent

Levine, M. (2018). *The blood unicorn Theranos was just a fairy tale*. Bloomberg. https://www.bloomberg.com/opinion/articles/2018-03-14/theranos-misled-investors-and-consumers-who-used-its-blood-test

Lewis, M. (2010). *The big short: Inside the doomsday machine*. W. W. Norton & Company.

Li, J., Kim, B., & Moore, J. *Investing in ElevenLabs*. a16z. https://a16z.com/announcement/investing-in-elevenlabs/

Li, X., & Mills, M. (2019). Vocal features: From voice identification to speech recognition by machine. *Technology and Culture, 60*(2), S129–S160.

Liberman, M., & Wayne, C. (2020). Human language technology. *AI Magazine, 41*(2), 22–35. https://doi.org/10.1609/aimag.v41i2.5297

Lighthill, J. (1973). Artificial intelligence: A general survey. *Artificial intelligence: A paper symposium*. Science Research Council.

Liu, T., Yuan, Z., Sun, J., Wang, J., Zheng, N., Tang, X., & Shum H. Y. (2011). Learning to detect a salient object. *IEEE Transactions on Pattern Analysis and Machine Intelligence, 33*(2), 353–367.

Livingstone, S. R., & Russo, F. A. (2018). The Ryerson Audio-Visual Database of Emotional Speech and Song (RAVDESS): A dynamic, multimodal set of facial and vocal expressions in North American English. *PLoS One, 13*(5), e0196391.

Lorenzo-Trueba, J., Yamagishi, J., Toda, T., Saito, D., Villavicencio, F., Kinnunen, T., & Ling, Z. (2018). *The voice conversion challenge 2018: Promoting development of parallel and nonparallel methods*. ArXiv. https://doi.org/10.48550/arXiv.1804.04262

Lorin, J., & Bloomberg (2024). *Harvard becomes latest Ivy league school to backtrack SAT admissions requirement.* https://fortune.com/2024/04/11/harvard-sat-admissions-requirement-ivy-league-colleges/

Luchanka, A. (2023). *The path to joining the club: How long it takes to become a unicorn.* Vention. https://ventionteams.com/blog/how-long-it-takes-to-become-a-unicorn

Mackenzie, A. (2017). *Machine learners: Archaeology of a data practice.* MIT Press.

Malabou, C. (2021). *Morphing intelligence: From IQ measurement to artificial brains* (C. Shread, Trans.). Columbia University Press. (Original work published 2019)

Marcus, G. (2025a). *Seven replies to the viral Apple reasoning paper—And why they fall short.* Marcus on AI. https://garymarcus.substack.com/p/seven-replies-to-the-viral-apple

Marcus, G. (2025b). *Five quick updates about that Apple reasoning paper that people can't stop talking about.* Marcus on AI. https://garymarcus.substack.com/p/five-quick-updates-about-that-apple

Martin, O. J. F., Kotsia, I., Macq, B., & Pitas, I. (2006). The eNTERFACE'05 audio-visual emotion database. In *22nd International Conference on Data Engineering Workshops (ICDEW '06)* (p. 8). IEEE.

Mauss, M. (1973). Techniques of the body (B. Brewster, Trans.). *Economy and Society, 2,* 70–88. (Original work published 1936)

McKay, A. (2015). *The big short* [Film]. Paramount Pictures.

McKeown, G., Valstar, M. F., Cowie, R., & Pantic, M. (2010). The SEMAINE corpus of emotionally coloured character interactions. In *2010 IEEE International Conference on Multimedia and Expo (ICMO '10)* (pp. 1079–1084). IEEE.

McLaren, M., Ferrer, L., Castan, D., & Lawson, A. (2016). The Speakers in the Wild (SITW) speaker recognition database. In *Proceedings of the 17th Annual Conference of the International Speech Communication Association, INTERSPEECH 2016* (pp. 818–822). ISCA. http://dx.doi.org/10.21437/Interspeech.2016-1129

McStay, A. (2018). *Emotional AI: The rise of empathic media.* Sage.

Megill, A. (Ed.). (1994). *Rethinking objectivity: Post contemporary interventions.* Duke University Press.

Mendon-Plasek, A. (2021). Mechanized significance and machine learning: Why it became thinkable and preferable to teach machines to judge the world. In J. Roberge and M. Castelle (Eds.), *The cultural life of machine learning* (pp. 31–78). Palgrave Macmillan.

"Mental Health and Wellness." (2022). https://www.voicesense.com/usecases/mental-health-and-wellness

Miller, G. A. (2003). The cognitive revolution: A historical perspective. *Trends in Cognitive Sciences, 7*(3), 141–144. https://doi.org/10.1016/S1364-6613(03)00029-9

Mills, M. (2010). Deaf jam: From inscription to reproduction to information. *Social Text, 28*(1), 35–58.

Mills, M. (2011). Hearing aids and the history of electronics miniaturization. *IEEE Annals of the History of Computing, 33*(2), 24–45.

Mills, M. (2015). Technology. In R. Adams, B. Reiss, & D. Serlin (Eds.), *Keywords for Disability Studies* (pp. 176–180). New York University Press.

Mims, C. (2021). *Why artificial intelligence isn't intelligent.* https://www.wsj.com/tech/ai/why-artificial-intelligence-isnt-intelligent-11627704050?gaa_at=eafs&gaa_n=AWEtsqd3PhHt03LJUUxHaXxA0hvcOG3Ffav29xbpNCYzok-BeX5R_ft-Snj2iY6jb64%3D&gaa_ts=6944532b&gaa_sig=vMMOhs4Yv7h3STJ2QrNjoJmz2DXDNBsuUcgR01yW8u3uhHmtbn3an6ukMe5yhhksa8mKR3pUShmsd3ykdWy7_w%3D%3D

Morrison, G. S., & Thompson, W. C. (2017). Assessing the admissibility of a new generation of forensic voice comparison testimony. *The Columbia Science & Technology Law Review, 18*(2), 326–434. https://ssrn.com/abstract=2883767

Muldoon, J., Cant, C., Wu, B., & Graham, M. (2024). A typology of artificial intelligence data work. *Big Data & Society, 11*(1). https://doi.org/10.1177/20539517241232632

Mulvin, D. (2021). *Proxies: The cultural work of standing in.* MIT Press.

Munn, L. (2020). *Logic of feeling: Technology's quest to capitalize emotion.* Rowman & Littlefield.

Murray, I. R., & Arnott, J. L. (1993). Toward the simulation of emotion in synthetic speech: A review of the literature on human vocal emotion. *The Journal of the Acoustical Society of America, 93*(2), 1097–1108.

MyHealthyApple. (2022). *Amazon Halo tone analysis: A beginner's user guide and review.* https://www.myhealthyapple.com/amazon-halo-tone-analysis-user-guide-review/

Nagrani, A., Chung, J. S., & Zisserman, A. (2017). *VoxCeleb: A large-scale speaker identification dataset.* ArXiv. https://doi.org/10.48550/arXiv.1706.08612

Nagrani, A., Chung, J. S., Xie, W., & Zisserman, A. (2020). Voxceleb: Large-scale speaker verification in the wild. *Computer Speech & Language, 60,* 1–15. https://doi.org/10.1016/j.csl.2019.101027

Nagy, J. (2022). Autism and the making of emotion AI: Disability as resource for surveillance capitalism. *New Media & Society, 26*(7), 3989–4007.

Nakamura, K. (2019). My algorithms have determined you're not human: AI-ML, reverse Turing-tests, and the disability experience. In *Proceedings of the 21st International*

ACM SIGACCESS Conference on Computers and Accessibility (ASSETS '19) (pp. 1–2). Association for Computing Machinery.

Narayanan, A., & Kapoor, S. (2024). *AI snake oil: What artificial intelligence can do, what it can't, and how to tell the difference*. Princeton University Press.

National Research Council. (1966). *Language and machines: Computers and translations in linguistics*. National Academies Press. https://doi.org/10.17226/9547

National Research Council. (1979). *On the theory and practice of voice identification*. National Academies Press. https://doi.org/10.17226/19814

Neisser, U. (1963). The imitation of man by machine. *Science, 139*(3551), 193–197.

Neisser, U. (1967). *Cognitive psychology*. Prentice Hall.

Netzorg, R., Yu, B., Guzman, A., Wu, P., McNulty, L., & Anumanchipalli, G. K. (2024). Towards an interpretable representation of speaker identity via perceptual voice qualities. In *ICASSP 2024–2024 IEEE International Conference on Acoustics, Speech and Signal Processing (ICASSP)* (pp. 12391–12395). Institute of Electrical and Electronics Engineers. https://doi.org/10.1109/ICASSP48485.2024.10446197

Newell, A., & Simon, H. (1956). The logic theory machine: A complex information processing system. *IRE Transactions on Information Theory, 2*(3), 61–79.

Newlands, G. (2021). Lifting the curtain: Strategic visibility of human labour in AI-as-a-service. *Big Data & Society, 8*(1). https://doi.org/10.1177/205395172110 16026

Nicholas, T. (2019). *VC: An American history*. Harvard University Press.

Niland, T. M., Duling, J., Allen, V., & Panther, E. (1971). Student counselors' perceptions of videotaping. *Counselor Education and Supervision, 11*(2), 97–101. https://doi .org/10.1002/j.1556-6978.1971.tb01491.x

Nogueiras, A., Moreno, A., Bonafonte, A., & Mariño, J. B. (2001). Speech emotion recognition using hidden Markov models. In *Eurospeech 2001: Seventh European Conference on Speech Communication and Technology* (pp. 1–14). International Speech Communication Association.

Nolan, C. (Director). (2014). *Interstellar* [Film]. Legendary Pictures, Syncopy, & Lynda Obst Productions.

Nuance Communications. (2020). *Nuance Gatekeeper Engine* [Data sheet]. https:// www.nuance.com/content/dam/nuance/en_us/collateral/enterprise/data-sheet/ds -nuance-gatekeeper-engine-en-us.pdf

"Nuance Public Safety and Security Forensic Biometric Solutions." (n.d.). https:// www.nuance.com/omni-channel-customer-engagement/security/public-safety .html

Nwe, T. L., Foo, S. W., & De Silva, L. C. (2003). Speech emotion recognition using hidden Markov models. *Speech Communication, 41*(4), 603–623. https://doi.org/10.1016/S0167-6393(03)00099-2

O'Brien, M., Chan, K., & Beaty, T. (2024). *OpenAI looks to shift away from nonprofit roots and convert itself to for-profit company.* Associated Press. https://apnews.com/article/chatgpt-openai-sam-altman-nonprofit-859bff5c19845f51796244e0072e2dfb

OpenAI. (2023). *ChatGPT can now see, hear, and speak.* https://openai.com/blog/chatgpt-can-now-see-hear-and-speak

OpenAI. (2024). *GPT-4 technical report.* ArXiv. https://doi.org/10.48550/arXiv.2303.08774

Orr, W., & Crawford, K. (2024). The social construction of datasets: On the practices, processes, and challenges of dataset creation for machine learning. *New Media & Society, 26*(9), 4955–4972. https://doi.org/10.1177/14614448241251797

Orr, W., & Kang, E. B. (2024). AI as a sport: On the competitive epistemologies of benchmarking. In *Proceedings of the 2024 Conference on Fairness, Accountability, and Transparency (FAccT'24)* (pp. 1875–1884). Association for Computing Machinery. https://doi.org/10.1145/3630106.3659012

Pasler, J. (2004). The utility of musical instruments in the racial and colonial agendas of late nineteenth-century France. *Journal of the Royal Music Association 129*(1), 24–76.

Passi, S., & Sengers, P. (2020). Making data science systems work. *Big Data & Society, 7*(2), 2053951720939605.

Peachey, K. (2019). *HMRC forced to delete five million voice files.* BBC. https://www.bbc.com/news/business-48150575

People v. King, 266 Cal. App. 2d 437, 72 Cal. Rptr. 478 (Ct. App. 1968).

Perrigo, B. (2023). *OpenAI used Kenyan workers on less than $2 per hour to make ChatGPT less toxic.* TIME. https://time.com/6247678/openai-chatgpt-kenya-workers/

Perrigo, B. (2024). *Meta's AI chief Yann LeCun on AGI, open-source, and AI-risk.* TIME. https://time.com/6694432/yann-lecun-meta-ai-interview/

Picard, R. (1995). *Affective computing.* MIT Press.

Poirier, L. (2021). Reading datasets: Strategies for interpreting the politics of data signification. *Big Data & Society, 8*(2), 20539517211029322.

Polanyi, M. (1958). *Personal knowledge: Towards a post-critical philosophy.* University of Chicago Press.

Poria, S., Hazarika, D., Majumder, N., Naik, G., Cambria, E., & Mihalcea, R. (2019). *MELD: A multimodal multi-party dataset for emotion recognition in conversations.* ArXiv. https://doi.org/10.48550/arXiv.1810.02508

Porter, T. M. (1995). *Trust in numbers: The pursuit of objectivity in science and public life*. Princeton University Press.

Potter, R. K., Kopp, G. A., & Green, H. C. (1944). *Voice print identification procedure and information*. Bell Telephone Laboratory Annual Report.

Press, G. (2017). *AI is a mirror, not a master, says Tim O'Reilly*. Forbes. https://www.forbes.com/sites/gilpress/2022/11/28/ai-is-a-mirror-not-a-master-says-tim-oreilly/

Putnam, H. (1988). *Representation and reality*. MIT Press.

Raji, D., Denton, E., Bender, E. M., Hanna, A., & Paullada, A. (2021). AI and the everything in the whole wide world benchmark. In *Proceedings of the Neural Information Processing Systems Track on Datasets and Benchmarks 1 (NeurIPS Datasets and Benchmarks 2021)* (pp. 1–17).

Raley, R., & Rhee, J. (2023). Critical AI: A field in formation. *American Literature, 95*(2), 185–204.

Ramati, I. (2024). Algorithmic ventriloquism: The contested state of voice in AI speech generators. *Social Media + Society, 10*(1).

Rancière, J. (2004). *The politics of aesthetics: The distribution of the sensible* (G. Rockhill, Trans.). Continuum. (Original work published 2000)

"Reduce Employee Churn." (n.d.). https://cogitocorp.com/solutions/reduce-employee-churn/

Rescorla, M. (2020). The computational theory of mind. In E. N. Zalta (Ed.), *The Stanford encyclopedia of philosophy (Fall 2020 edition)*. https://plato.stanford.edu/entries/computational-mind/

Reich, A. R., Moll, K. L., & Curtis, J. F. (1976). Effects of selected vocal disguises upon spectrographic speaker identification. *The Journal of the Acoustical Society of America, 60*(4), 919–925. https://doi.org/10.1121/1.381173

Reynolds, D. A., Quatieri, T. F., & Dunn, R. B. (2000). Speaker verification using adapted Gaussian mixture models. *Digital Signal Processing, 10*(1–3), 19–41. https://doi.org/10.1006/dspr.1999.0361

Rice, T. (2012). Sounding bodies: Medical students and the acquisition of stethoscopic perspectives. In T. Pinch, & K. Bijsterveld (Eds.), *The Oxford handbook of sound studies* (pp. 298–319). Oxford University Press.

Rice, T. (2015). Listening. In D. Novak & M. Sakakeeny (Eds.), *Keywords in sound* (pp. 99–111). Duke University Press.

Robbie [@Robbie92_]. (2023, June 30). *With the rice of AI voice cloning, voice actors are being abused by the modding communities. As a member of* [Image attached] [Tweet]. X. https://x.com/Robbie92_/status/1674944276109688833

Roberge, J., & Castelle, M. (Eds.). (2021). *The cultural life of machine learning: An incursion into critical AI studies*. Palgrave Macmillan.

Roberts, R. R., Jr., & Renzaglia, G. A. (1965). The influence of tape recording on counseling. *Journal of Counseling Psychology, 12*(1), 10. https://doi.org/10.1037/h0021936

Roemmich, K., Rosenberg, T., Fan, S., & Andalibi, N. (2023). Values in emotion artificial intelligence hiring services: Technosolutions to organizational problems. In *Proceedings of the ACM Conference in Human Computer Interaction (PACMHCI '23)*. Association for Computing Machinery.

Roff, C. (2024). *Everything you need to know about the Theranos scandal*. Pan Macmillan. https://www.panmacmillan.com/blogs/literary/theranos-elizabeth-holmes-john-carreyrou

Rohrmann, S., Bechtoldt, M. N., Hopp, H., Hodapp, V., & Zapf, D. (2011). Psychophysiological effects of emotional display rules and the moderating role of trait anger in a simulated call center. *Anxiety, Stress & Coping, 24*(4), 421–438.

Romero, A. (2021). *What would the world look like if AI wasn't called AI?* https://towardsdatascience.com/what-would-the-world-look-like-if-ai-wasnt-called-ai-bfb5ae35e68a

Rose, J. (2022). *This tool lets anyone see the bias in AI image generators*. https://www.vice.com/en/article/bvm35w/this-tool-lets-anyone-see-the-bias-in-ai-image-generators

Rose, S. (2017). *No right to be idle: The invention of disability, 1840s–1930s*. University of North Carolina Press.

Rossi, A. (Director). (2022). *The Andy Warhol Diaries* [TV Series]. Abstract Dakota Group Ltd. & Submarine Deluxe.

Rothenberg, E. (2024). *Should the SAT still matter after all these years? Why some colleges are bringing it back*. CNN. https://www.cnn.com/2024/04/21/business/sat-college-requirement-2024-test/index.html#:~:text=A%20strong%20emphasis%20on%20high,2023%20from%20market%20researcher%20Technavio

Rubin, D. L. (1992). Nonlanguage factors affecting undergraduates' judgments of nonnative English-speaking teaching assistants. *Research in Higher Education, 33*, 511–531.

Russakovsky, O., Deng, J., Su, H., Krause, J., Satheesh, S., Ma, S., Huang, Z., Karpathy, A., Khosla, A., Bernstein, M., Berg, A. C., & Fei-Fei, L. (2015). ImageNet Large Scale Visual Recognition Challenge. *International Journal of Computer Vision, 115*(3), 211–252. https://doi.org/10.1007/s11263-015-0816-y

Russell, J. A. (1991). Culture and the categorization of emotions. *Psychological Bulletin, 110*(1), 145–172.

Russell, J. A. (2003). Core affect and the psychological construction of emotion. *Psychological Review, 110*(1), 145–172.

Sadowski, J. (2025). Machine's eye view: Postmodern data science and the politics of ground truth. *Science, Technology, & Human Values*. Advance online publication. https://doi.org/10.1177/01622439251331138

Santos, D. (2023). The cyborg's prosody, or speech AI and the displacement of feeling. *Sounding Out!* https://soundstudiesblog.com/2023/07/03/25303/

Schafer, R. M. (1993). *The soundscape: Our sonic environment and the tuning of the world*. Destiny Books. (Original work published 1977)

Schellmann, H. (2024). *The algorithm: How AI decides who gets hired, monitored, promoted, and fired and why we need to fight back now*. Da Capo Press.

Scherer, K. R. (2009). The dynamic architecture of emotion: Evidence for the component process model. *Cognition and Emotion, 23*, 1307–1351.

Scherer, K. R., Johnstone, T., & Klasmeyer, G. (2003). Vocal expression of emotion. In R. J. Davidson, K. R. Scherer, & H. Goldsmith (Eds.), *Handbook of affective sciences* (pp. 433–456). Oxford University Press.

Schuller, B. W. (2018). Speech emotion recognition: Two decades in a nutshell, benchmarks, and ongoing trends. *Communications of the ACM, 61*(5), 90–99.

Schuller, B., Rigoll, G., & Lang, M. (2003). Hidden Markov model-based speech emotion recognition. In *Proceedings of the 2003 IEEE International Conference on Acoustics, Speech, and Signal Processing (ICASSP'03)* (Vol. 2, pp. ii–1). IEEE.

Schuller, B., Rigoll, G., & Lang, M. (2004). Speech emotion recognition combining acoustic features and linguistic information in a hybrid support vector machine-belief network architecture. In *2004 IEEE International Conference on Acoustics, Speech, and Signal Processing* (Vol. 1, pp. i–577). IEEE.

Schulz, J., & Dincer, M. (2022). Amici brief of science, legal, and technology scholars in Renderos et al. v. Clearview AI, Inc. et al., No. RG21096898 (Superior Ct. Alameda County). NYU School of Law, Public Law Research Paper No. 23–18. http://dx.doi.org/10.2139/ssrn.4238870

Seaver, N. (2019). Knowing algorithms. In J. Vertesi & D. Ribes (Eds.), *digitalSTS: A field guide for science and technology studies* (pp. 412–422). Princeton University Press.

Securus Technologies. (2018). *Now available: Securus' Investigator Pro™ with searchable voice capability*. https://www.prnewswire.com/news-releases/now-available-securus-investigator-pro-with-searchable-voice-capability-300376567.html

Selfridge, O. G., & Neisser, U. (1960). Pattern recognition by machine. *Scientific American, 203*(2), 60–69.

Semel, B. M. (2022). Listening like a computer: Attentional tensions and mechanized care in psychiatric digital phenotyping. *Science, Technology, & Human Values, 47*(2), 266–290.

Serra-Garcia, M., & Gneezy, U. (2021). Nonreplicable publications are cited more than replicable ones. *Science Advances, 7*(21). https://doi.org/10.1126/sciadv.abd1705

Shalev-Shwartz, S., & Ben-David, S. (2014). *Understanding machine learning: From theory to algorithms.* Cambridge University Press.

Shapin, S., & Schaffer, S. (1985). *Leviathan and the air-pump: Hobbes, Boyle and the experimental life.* Princeton University Press.

Shaw, G. B. (1994). *Pygmalion.* Dover Publications. (Original work published 1916)

Shojaee, P., Mirzadeh, I., Alizadeh, K., Horton, M., Bengio, S., & Farajtabar, M. (2025). *The illusion of thinking: Understanding the strengths and limitations of reasoning models via the lens of problem complexity.* Apple. https://machinelearning.apple.com/research/illusion-of-thinking

Siebers, T. (2008). *Disability theory.* University of Michigan Press.

Simonite, T. (2018). This call may be monitored for tone and emotion. *Wired.* https://www.wired.com/story/this-call-may-be-monitored-for-tone-and-emotion/

Skitka, L. J., Mosier, K. L., & Burdick, M. (1999). Does automation bias decision making? *International Journal of Human–Computer Studies, 51*(5), 991–1006.

Spearman, C. (1904). "General intelligence," objectively determined and measured. *The American Journal of Psychology, 15*(2), 201–293. https://doi.org/10.2307/1412107

Srećković, S., Berber, A., & Filipović, N. (2021). The automated Laplacean demon: How ML challenges our views on prediction and explanation. *Minds and Machines, 32,* 159–183. https://doi.org/10.1007/s11023-021-09575-6

Staniszewski, M. (2024). *ElevenLabs releases new voice AI products and raises $80M series B.* ElevenLabs. https://elevenlabs.io/blog/series-b/

Stanovich, K. (2009). *What intelligence tests miss: The psychology of rational thought.* Yale University Press.

Star, S. L. (1983). Simplification in scientific work: An example from neuroscience research. *Social Studies of Science, 13*(2), 205–228.

Stark, L., & Hoey, J. (2021). The ethics of emotion in artificial intelligence systems. In *Proceedings of the 2021 ACM Conference on Fairness, Accountability, and Transparency (FAccT '21)* (pp. 782–793). Association for Computing Machinery.

Sterne, J. (2003). *The audible past: Cultural origins of sound reproduction.* Duke University Press.

Sterne, J. (2012). Sonic imaginations. In J. Sterne (Ed.), *The sound studies reader* (pp. 1–18). Routledge.

Sterne, J. (2022). Is machine listening listening? *communication+1, 9*(1). https://doi.org/10.7275/zeqh-eg38

Stoever, J. L. (2017). *The sonic color line: Race and the cultural politics of listening* (Vol. 17). NYU Press.

"Students Measure Fear by a Pupilometer, Kick Subject's Shins to Experiment on Anger." (1925, November 24). *The New York Times*, sec. 27:4.

Suchman, L. (2023). The uncontroversial "thingness" of AI. *Big Data & Society, 10*(2). https://doi.org/10.1177/20539517231206794

Sulpizio, S., Fasoli, F., Maass, A., Paladino, M. P., Vespignani, F., Eyssel, F., & Bentler, D. (2015). The sound of voice: Voice-based categorization of speakers' sexual orientation within and across languages. *PLoS One, 10*(7), e0128882. https://doi.org/10.1371/journal.pone.0128882

Tanney, M. F., & Gelso, C. J. (1972). Effect of recording on clients. *Journal of Counseling Psychology, 19*(4), 349–350. https://doi.org/10.1037/h0033077

Terminator Wiki. (n.d.). SkyNet. https://terminator.fandom.com/wiki/Skynet

Thier, J. (2024). *AI is moving too fast to keep pace for 4 in 5 workers*. Fortune. https://fortune.com/2024/02/06/ai-acceleration-too-fast-workers-springboard/

Thomas, F., & Johnston, O. (1981). *Disney animation: The illusion of life*. Abbeville Press.

Thurstone, L. L. (1935). *The vectors of mind*. University of Chicago Press.

Thurstone, L. L. (1940). Current issues in factor analysis. *Psychological bulletin, 37*, 189–236.

Toda, T., Chen, L., Saito, D., Villavicencio, F., Wester, M., Wu, Z., & Yamagishi, J. (2016). The voice conversion challenge 2016. In *Proceedings of Interspeech 2016* (pp. 1632–1636). International Speech Communication Association. http://dx.doi.org/10.21437/Interspeech.2016-1066

Tomkins, S. (1962). *Affect imagery consciousness: Volume I: The positive affects*. Springer.

Tomkins, S. (1963). *Affect imagery consciousness: Volume 2: The negative affects*. Springer.

Toosi, A., Bottino, A. G., & Saboury, B. (2021). A brief history of AI: How to prevent another winter (a critical review). *PET Clinics, 16*(4), 449–469. https://doi.org/10.1016/j.cpet.2021.07.001

Tosi, O., Oyer, H., Lashbrook, W., Pedrey, C., Nicol, J., & Nash, E. (1972). Experiment on voice identification. *The Journal of the Acoustical Society of America, 51*(6B), 2030–2043. https://doi.org/10.1121/1.1913064

Tubaro, P., Casilli, A. A., & Coville, M. (2020). The trainer, the verifier, the imitator: Three ways in which human platform workers support artificial intelligence. *Big Data & Society, 7*(1). https://doi.org/10.1177/2053951720919776

Turing, A. M. (1950). Computing machinery and intelligence. *Mind, LIX*(26), 433–460.

Turnbridge, L. (2016). Scarlett Johansson's body and the materialization of voice. *Twentieth Century Music, 13*(1), 131–152.

Van Rossem, W., & Pelizza, A. (2022). The ontology explorer: A method to make visible data infrastructures for population management. *Big Data & Society, 9*(1), 20539517221104087.

Vaswani, A., Shazeer, N., Parmar, N., Uszkoreit, J., Jones, L., Gomez, A. N., Kaiser, L., & Polosukhin, I. (2017). Attention is all you need. In *Proceedings of the 31st International Conference on Neural Information Processing Systems (NIPS'17)* (pp. 6000–6010). Curran Associates.

Voicesense. (n.d.). https://www.voicesense.com/home

Walther, C. C. (2025). *Intelligence illusion: What Apple's AI study reveals about reasoning.* Forbes. https://www.forbes.com/sites/corneliawalther/2025/06/09/intelligence-illusion-what-apples-ai-study-reveals-about-reasoning/

Wang, A., Singh, A., Michael, J., Hill, F., Levy, O., & Bowman, S. R. (2019). *GLUE: A multi task benchmark and analysis platform for natural language understanding.* ArXiv. https://doi.org/10.48550/arXiv.1804.07461

Wang, D., Prabhat, S., & Sambasivan, N. (2022, April 29–May 5). Whose AI dream? In search of the aspiration in data annotation. In *CHI Conference on Human Factors in Computing Systems (CHI '22)*, New Orleans, LA, United States.

Weidman, A. (2015). Voice. In D. Novak & M. Sakakeeny (Eds.). *Keywords in sound* (pp. 232–245). Duke University Press.

Whitaker, M, Alper, M., Bennett, C. L., Hendren, S., Kaziunas, L., Mills, M., Ringel Morris, M., Rankin, J., Rogers, E., Salas, M., & Myers West, S. (2019). *Disability, bias, and AI.* https://ainowinstitute.org/disabilitybiasai-2019.pdf

Williams, A., Miceli, M., & Gebru, T. (2022). *The exploited labor behind artificial intelligence.* https://www.noemamag.com/the-exploited-labor-behind-artificial-intelligence/

Wong, M. (2023). *AI doomerism is a decoy.* The Atlantic. https://www.theatlantic.com/technology/archive/2023/06/ai-regulation-sam-altman-bill-gates/674278/

Wu, X., & Zhang, X. (2016a). *Automated inference on criminality using face images.* ArXiv. https://doi.org/10.48550/1611.04135v1

Wu, X., & Zhang, X. (2016b). *Responses to critiques on machine learning of criminality perceptions (addendum of arXiv: 1611.04135).* ArXiv. https://doi.org/10.48550/arXiv.1611.04135

Yenawine, G., & Arbuckle, D. S. (1971). Study of the use of videotape and audiotape as techniques in counselor education. *Journal of Counseling Psychology, 18*(1), 1–6. https://doi.org/10.1037/h0030394

Zhao, Y., Huang, W. C., Tian, X., Yamagishi, J., Das, R. K., Kinnunen, T., Ling, Z., & Toda, T. (2020). *Voice conversion challenge 2020: Intra-lingual semi-parallel and cross-lingual voice conversion*. ArXiv. https://doi.org/10.48550/arXiv.2008.12527

Zuboff, S. (2019). *The age of surveillance capitalism: The fight for a human future at the new frontier of power*. PublicAffairs.

Index

NOTE: Locators containing *f* indicate figures; locators containing *t* indicate tables.

Publisher contact:
The MIT Press
Massachusetts Institute of Technology
77 Massachusetts Avenue, Cambridge, MA 02139
mitpress.mit.edu

EU Authorised Representative:
Easy Access System Europe, Mustamäe tee 50,
10621 Tallinn, Estonia
gpsr.requests@easproject.com

Printed by Integrated Books International,
United States of America